Alligators of Texas

TWENTY-NINE

Gulf Coast Books
Sponsored by
Texas A&M University–Corpus Christi
John W. Tunnell Jr., General Editor

*A list of titles in this series is available
at the end of the book.*

Alligators of Texas

Louise Hayes

Photographs by Philippe Henry

Texas A&M University Press • College Station

This paper meets the requirements of ANSI/NISO
Z39.48–1992 (Permanence of Paper).
Binding materials have been
chosen for durability.
Manufactured in China
by Everbest Printing Co.
through FCI Print Group
∞

LIBRARY OF CONGRESS CATALOGING-IN-PUBLICATION DATA

Names: Hayes, Louise, 1957– author. | Henry,
    Philippe, 1954– photographer.
Title: Alligators of Texas / Louise Hayes ; photo-
    graphs by Philippe Henry.
Other titles: Gulf Coast books ; no. 29.
Description: First edition. | College Station : Texas
    A&M University Press, [2016] | Series: Gulf Coast
    books ; twenty-nine | Includes bibliographical
    references and index.
Identifiers: LCCN 2015044743| ISBN 9781623493875
    (flexbound (with flaps) ; alk. paper) |
    ISBN 9781623493882 (ebook)
Subjects: LCSH: American alligator—Texas.
Classification: LCC QL666.C925 H38 2016 | DDC
    597.98/409764—dc23 LC record available at
    http://lccn.loc.gov/2015044743

To my parents,
Louis and Renata Hayes,
who have always been there for me
and have supported any endeavor I took on

LOUISE HAYES

In memory of our dear friend
and field companion Bill Howell

LOUISE HAYES & PHILIPPE HENRY

Contents

Preface

When I was a young child, my first experiences and memories with crocodilians involved crocodiles, one living and the other an English handmade wooden puppet. The living crocodilian, a slender-snouted crocodile residing at the St. Louis Zoo, had a sign in its enclosure stating that he was the zoo's oldest resident. Every time I visited the zoo, the crocodile would bask on the "land" area of his exhibit adjacent to the sign. He always had his mouth gaping. I knew that the mouth being open was not by accident. I just did not understand its significance or its relationship to his basking in the sun under a skylight. On one visit, he was underwater in his pond located in front of the basking area. I wondered why he was in the water rather than on land. Those questions were answered years later when I learned how crocodilians thermoregulate. The crocodile, which arrived at the zoo as a juvenile in 1933, lived until 1989, when he succumbed to peritonitis.

The crocodile puppet belonged to my English aunt through marriage (a World War II war bride), who was a puppeteer. It was the only hand puppet in her "Punch and Judy" show. The crocodile swallowed sausages, and its jaws popped loudly and menacingly together when it did so. My older cousin once frightened me by chasing me around with the puppet when I was very young. When I was a few years older, my aunt let my sister and me play

with her puppets, after presenting a puppet show for us and some friends, so I solved the mystery of how the crocodile swallowed the sausages and admired its craftsmanship. I note that my aunt's father was a veterinarian in the British military and killed a large Nile crocodile while in Africa. My aunt had a photograph hanging in her hallway of him posing with the impressive animal, and I made it a point to look at the photograph every time I visited.

Alligators played an important role in how I ended up in Texas. I started working at the Henson Robinson Zoo in Springfield, Illinois, when I was 16 years old. I worked with a variety of reptiles, including alligators and caiman, as the zoo director was a herpetologist. After working there for three years, I went off to college. When I returned home several years later to finish writing my master's thesis, I worked at the zoo for several months. My goal after finishing my degree was to get a job working with reptiles at a major zoo. Like crocodilians, I love warm weather and wanted to relocate to someplace warm. I decided also that I wanted to live in a state that had alligators.

Lawrence Agnew, director of the Henson Robinson Zoo, knew the director of the Houston Zoo, John Werler, who also happened to be a herpetologist, and helped me get a job there. After working at the Houston Zoo for three and a half years, I decided to pursue a doctorate and conduct research on alligators in the wild in Texas for my dissertation. I contacted "Doc" Jim Dixon in the Wildlife Department at Texas A&M University. He had conducted research on alligators at the J. D. Murphree Wildlife Management Area and had also done research on crocodilians in the Neotropics. I went on to research alligators and complete a doctorate under his direction.

It was a privilege to work for and with two of the great Texas herpetologists, who later coauthored the books *Texas Snakes: Identification, Distribution, and Natural History* and *Texas Snakes: A Field Guide.* I have fond memories of having John and his wife, Ingrid, over for dinner. After earning my doctorate, I continued to interact with Doc and his wife, Mary, at Texas Herpetological Society meetings. I also visited with Doc when I came to College Station at the TAMU Cooperative Wildlife Research Lab, and I belonged to a Red Hat Society luncheon group with Mary, which was made up largely of biology and wildlife faculty members or the wives of faculty members. We lost John and Ingrid a number of years ago, and Doc just recently. They are missed by many.

Philippe Henry contacted me after reading in a Crocodile Specialist Group newsletter that an alligator I had tagged as a juvenile 15 years earlier was recaptured near the original capture site at Brazos Bend State Park. He asked if he could come to Texas and accompany me in the field at Brazos Bend. I agreed, and that was the beginning of our friendship and collaboration on a few alligator articles for European publications. Philippe suggested that we publish a book together, with Philippe providing the photographs and me writing the text. I contacted

Shannon Davies at Texas A&M University Press, told her what I had in mind for the book, and showed her Philippe's photographs. She was very receptive to acquiring the book.

Originally, only a very limited number of photographs could be included due to production costs. The earliest version of the text was written around the chosen photographs. The book then began to take on a life of its own as I came to the realization of what was important to include. Colleagues also made a big impact on the evolution of the book, both in general and in very specific terms, by suggesting what needed to be included. Complications caused the book to take longer to complete than expected but eventually allowed many other photographs to be included. A few vintage postcards and photographs are also included, particularly of the El Paso Plaza. My good friend and biologist Lisa McDonald contributed illustrations and photographs.

The book is meant for a wide audience yet is also designed for individuals with specialized interests, including wildlife biologists, game wardens, animal control personnel, land managers, alligator farmers, and concerned citizens dealing with alligator issues. As a result, this may not be a book that everyone wants to read from cover to cover. Some may want to enjoy Philippe's photographs and their captions or read those parts of the book that are of particular interest. The e-mail address (TexasAlligators@cs.com) and website (www.AlligatorsInTexas.com) for this book are a means for readers to share alligator stories, their experiences viewing alligators at zoos or in the wild, and anything else involving Texas alligators.

About the Photography

By spending a lot of time with the animal species that I photograph, I can learn about their behavior and life cycles. For five years, I spent weeks with the same alligators in Fort Bend County. Sometimes I could get very close and was able to film intimate moments. Every alligator has a different personality. Some will let you get close without displaying any aggressive behavior. Others will not.

Spending a lot of time with the animal species I want to film or photograph gives me the opportunity to watch the very small details. In spring 2004, I spent a lot of time with a pair of alligators during the courtship period. They allowed me to set my tripod and camera as close as 30 feet from the spot they had chosen for basking together after a short period of bellowing. Then I filmed the pair together with the young alligators that hatched in September 2004.

There is no mystery. The key to successful photography is being able to live with and stay around the animals that you study for a long period of time in order for them to accept you. After a while, you become like the animal itself. Your biological clock begins to run parallel with the inner clock of the animal, and you become part of its natural world. You can even feel what they feel. Often you can anticipate their actions and reactions. When you are close to this animal, there is a kind of communication that is unspoken but is a conveyance of feelings.

—Philippe Henry
philippe_henry@hotmail.com
www.philippe-henry.com

What Is an Alligator?

*The name alligator has been applied by the British settlers in the Southern
States to a species of reptile resembling in many respects, the crocodile
of Egypt. By some etymologists the name is supposed to be derived from
the Spanish term "el lagarto" (the lizard), while others assert that it is a
corruption or modification of the Indian word "legateer."*

—Captain Flack, *A Hunter's Experiences in the Southern States of America*

What is an alligator? That particular question
can be answered in terms of how it is related
to other organisms (taxonomy) in its hierarchy
from more primitive to increasingly advanced
features or in terms of how it is related to its
environment through adaptations.

Systematics and Taxonomy

The taxonomy, or naming system, of living
organisms is based on systematics, the study of
relationships among organisms based on ances-
try. See table 1.1 for the taxonomy of the American alligator.

The term "American alligator" is a common name, and the scientific name (consisting
of genus and species), *Alligator mississippiensis*, is always italicized (or underlined if writing

by hand). A species is a group of actually or
potentially breeding populations that is repro-
ductively isolated from other such groups. This
means that under normal circumstances a species cannot breed outside its group.

The American alligator is found only in
the United States and ranges throughout 10
states: Texas, Oklahoma, Arkansas, Louisiana,
Mississippi, Alabama, Georgia, Florida, North
Carolina, and South Carolina. The American
crocodile, the only other native crocodilian in
the United States, is restricted to extreme south
Florida and the Florida Keys. It also widely
occurs outside the United States in the Greater
Antilles (Cuba, Jamaica, Haiti, and the Domini-
can Republic) and from central Mexico through
parts of Central America to northern South
America. The spectacled caiman (*Caiman*

Table 1.1. Taxonomy of the American alligator

Kingdom	Phylum	Class	Order	Family	Genus	Species
Animalia	Chordata	Reptilia	Crocodylia	Crocodylidae	Alligator	mississippiensis

crocodilus) has been introduced to southern Florida through the pet trade, either from accidental escapees, or those that were deliberately released into the wild.

There is only one other living member of the genus *Alligator*, the Chinese alligator (*A. sinensis*), from the Yangtze River Valley in China. These two alligator species are the only temperate-climate crocodilians in existence today. The other crocodilian species are confined to the tropics.

There are more than 20 species of crocodilians. The number of species is in a state of flux at present, as some of the crocodile species are now believed to consist of at least two distinct species. The reptilian order, Crocodylia, is divided into two families, the Crocodylidae and Gavialidae. The American alligator belongs to the subfamily Alligatorinae, which also includes the Chinese alligator and the caimans. The crocodiles (*Crocodylus sp.*) belong to the subfamily Crocodylinae. The gharial (*Gavialis gangeticus*) and false gharial (*Tomistoma schlegelii*) make up the family Gavialidae. The false gharial, which has a long, slender nose like that of the gharial, has been moved between the crocodile and gharial families and was even placed in a family of its own until recently. At present, based on DNA evidence, it is now placed in the family Gavialidae.

The alligator and crocodile subfamilies are easy to differentiate from each other based on the fourth tooth (from front to back) on each side of the lower jaw. In the alligators and caimans this tooth fits into a bony socket in the upper jaw when the jaws are closed, so the tooth is hidden. In the crocodiles there is an indentation on each side of the upper jaw to accommodate the fourth tooth when the jaw is closed. This gives crocodiles a rather "toothy" appearance when the jaw is closed. If viewed from above, the indentation is very noticeable. Side and top views of gharial, alligator, and crocodile skulls are depicted in figure 1.1.

Crocodilians (and humans) belong to the phylum Chordata, which contains some invertebrates (sea squirts and lancelets) and all of the vertebrates. Three characteristics tie this group together: a dorsal (on the back) hollow nerve cord (the spinal cord) instead of a ventral (on the belly) one; a stiff, rodlike notochord that serves as an internal skeleton at some stage of life; and pharyngeal gill slits that extend from the pharynx (throat cavity) to the outside of the animal at some stage of life. The subphylum Vertebrata consists of animals possessing a vertebral column (backbone). Other characteristics of the vertebrates are cephalization (a definite head with sense organs and a concentration of nerves), a closed circulatory system (closed blood vessels), a ventral heart, liver, kidneys, and an endocrine system that secretes hormones. The notochord and pharyngeal gill slits are found only in the embryonic

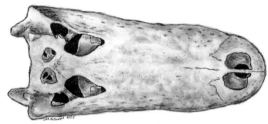

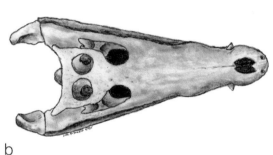

Figure 1.1.
(a) Side-view comparisons of skulls: gharial, alligator, and crocodile.
(b) Top-view comparison of skulls: gharial, alligator, and crocodile.
(Illustrations by Lisa McDonald)

a

b

stages of vertebrates. The notochord is surrounded or replaced by the vertebral column. The classes of vertebrates are Agnatha (jawless fish), Chondrichthyes (cartilaginous fish), Osteichthyes (bony fish), Amphibia (frogs, salamanders), Reptilia (crocodilians, turtles, lizards, snakes), Aves (birds), and Mammalia (humans). Table 1.2 lists the major characteristics of each vertebrate group.

The amphibians, reptiles, birds, and mammals are collectively known as tetrapods (tetra = four, pods = feet) and belong to the superclass Tetrapoda. These groups represent a change from aquatic to terrestrial life. Fins are needed to move in the water, whereas limbs are needed to move on land. The amphibians are considered to be a transitional group between an aquatic and terrestrial existence, and their name literally means "two lives." Amphibians range from being aquatic, to semi-aquatic, to terrestrial. However, even amphibians that live on land need water from the environment to reproduce. Many enter the water to breed and lay eggs and have an intermediate stage that lives in the water until it metamorphoses into an adult. Even amphibians that have direct development, in which eggs hatch on land into miniature adults without a larval stage, still

Table 1.2. Comparison of the vertebrate classes

Class	Representatives	Major characteristics
Agnatha	Lampreys, hagfish	Jawless fish with a suckerlike mouth; notochord present in adults; the eel-shaped lampreys are parasites and attach to other fish with their mouth to suck blood; hagfish feed on sick or dead fish or marine worms
Chondrichthyes	Sharks, skates, rays	Fish with cartilaginous skeletons; embedded toothlike scales; notochord replaced by vertebrae in adults; stiff fins; jaws present; gills for respiration; well-developed sense organs
Osteichthyes	Bass, perch, salmon, swordfish	"Bony" fish (i.e., have true skeleton); flat, bony scales in most species; flexible fins; jaws; majority have gills but a few possess lungs; many have a swim bladder (an air sac that controls buoyancy)
Amphibia	Anurans (frogs, toads), sala-manders, caecilians	Diverse, transitional group that is tied to water to keep the skin from drying out, or for respiration in the case of lungless salamanders, or to keep eggs of egg layers moist
Reptilia	Turtles, snakes, lizards, croco-dilians, tuatara	First vertebrate group to be entirely terrestrial; dry, scaly skin; lungs; amniote eggs if egg layers
Aves	Birds	Have many adaptations for flight, including feathers, front limbs mod-ified into wings, lightweight hollow bones, a beak to procure food, and a gizzard (an organ located near the stomach) instead of heavy teeth to grind food; endothermic (able to produce their own body heat to keep a constant body temperature)
Mammalia	Rodents, hoofed animals, carnivores, primates	Hair, mammary glands

require moisture from the environment to keep the eggs from desiccating (drying out).

The reptiles represent the first vertebrate group that is independent of the water. They still require water or moisture to facilitate the fertilization of eggs and sperm. They have an amniotic egg with a shell that takes "the water into the egg," rather than the egg requiring moisture from outside to remain hydrated. Other features that some reptiles, birds, and mammals have for a life on land include well-developed lungs enclosed by a protective rib cage; a well-developed brain; claws; limbs that can lift their body off the ground; dry, pro-tective scales that serve as a barrier from desic-cation and predators; and internal fertilization.

Characteristics of the Crocodilians

Crocodilians are known for having several fea-tures more advanced than those of other rep-tiles. They have thecodont teeth, which refers to the tooth attachment being in sockets. This reduces the chance of tooth loss when fighting or capturing prey.

The crocodilians have a complete four-chambered heart rather than the incomplete four-chambered heart of other reptiles, which has only a partial septum (wall) between the ventricles (lower chambers). Oxygenated blood coming from the lungs on the left side is separate from the oxygen-depleted blood returning from the body to the right side of the heart. However, the crocodilians do have a foramen of Panizzae (foramen = opening or hole) that allows some oxygenated blood from arteries of the left ventricle to mix with unoxygenated blood from arteries of the right ventricle. It was long thought that this mixing of oxygen-poor and oxygen-rich blood in reptiles was inefficient and that their incomplete four-chambered heart was an evolutionary step toward an efficient complete four-chambered heart as possessed by birds and mammals. This mixing (cardiac shunts) is now believed to be physiologically beneficial and varies with physiological state. Studies suggest that cardiac shunts are regulated and function in homeostasis (maintaining a steady state internally when conditions change). In crocodilians, it has been found that during diving, left ventricular pressure is high and causes a blood shunt through the foramen of Panizzae, resulting in all blood to the body being oxygenated.

The crocodilians possess a true cerebral cortex (the cerebrum is part of the forebrain, and the cortex is its outer layer), the portion of the brain involved with learning and memory. They possess complex social behavior and parental care usually associated more with birds than with other reptiles.

Adaptations to Habitat

An adaptation is something that helps an organism survive in its habitat, where a species lives. Adaptations can be physiological (functioning of cells, organs, and the entire body), morphological (form and structure), and behavioral. Crocodilians, as well as other reptiles, birds, and mammals, are primarily adapted for life on land. Fish are aquatic, and amphibians are a transitional group, ranging from aquatic to terrestrial. However, all living crocodilians are secondarily adapted to aquatic life. This aspect of their biology creates an interesting mix of external and physiological features. Secondary adaptations allow alligators and other crocodilians to live in both aquatic and terrestrial habitats. Note the "face plate" in figure 1.2. There is a similar arrangement on a hippo that also lives "where land and water meet." The crocodilians' eyes, ears, and nostrils look as though they were glued on top of the head. This portion of their head sticks out of the water when the rest of the body is submerged. The arrangement allows them to see, hear, and breathe in the level above the water. What happens when they put their head under the water's surface?

While underwater, crocodilians utilize a nictitating membrane, a third eyelid that is transparent and moves across from one side of the eye to the other (like a curtain) to waterproof and protect it (fig. 1.3). Most likely vision underwater is not clear. At the side of the crocodilian head, the slitlike ear openings are evident just behind the eyes (fig. 1.4). They close when the animal is underwater, as do the nostrils.

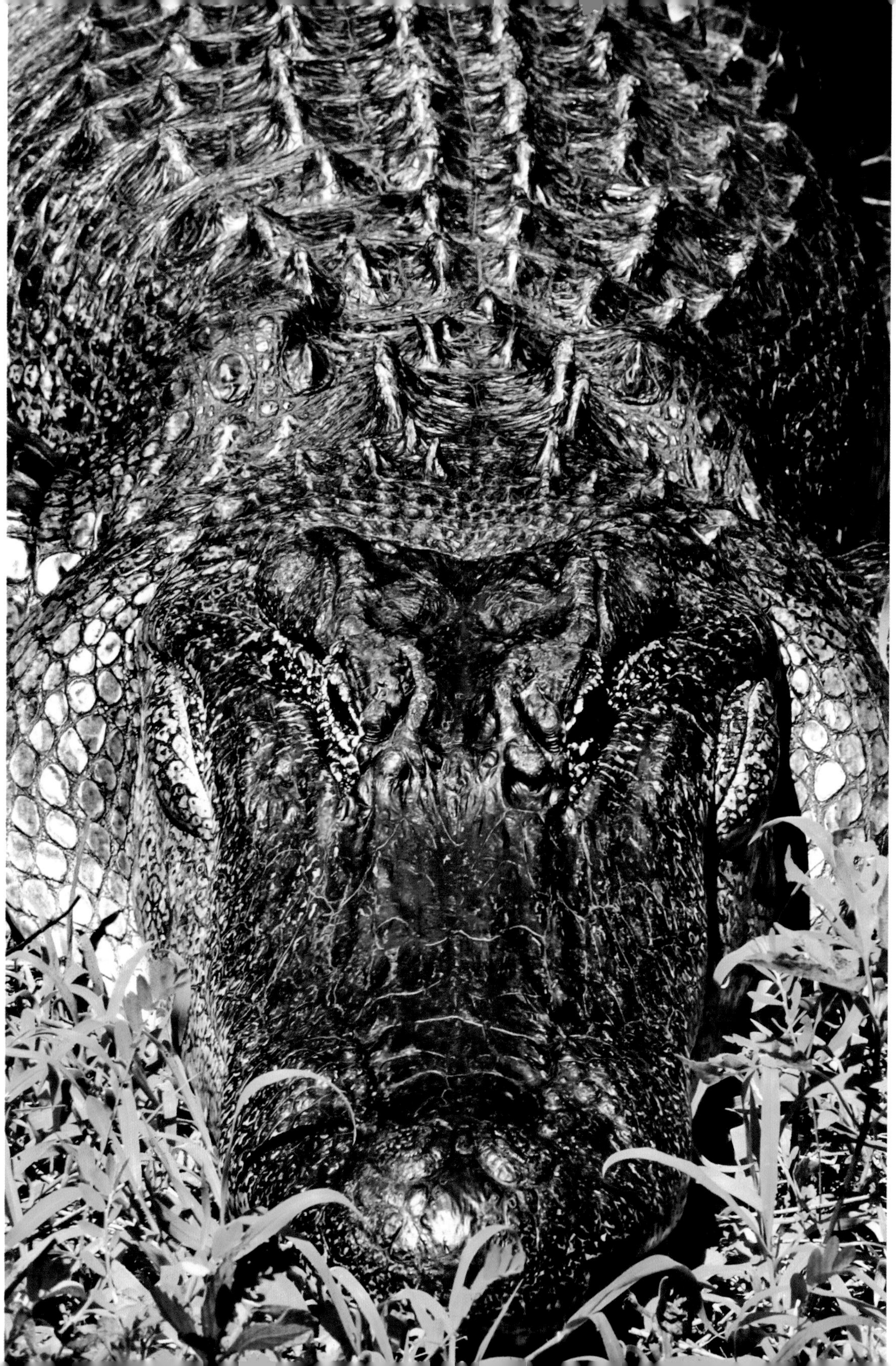

Figure 1.2. The "face plate" of an alligator: the parts of the head that stay above the waterline, allowing the animal to breathe, see, and hear.

The nostrils open into the nasal passages, which are kept separate from the mouth by the secondary palate. The nasal passages extend into the back of the throat, and a trachea or windpipe extends from the throat into the lungs. Unlike humans with fleshy lips, the crocodilian mouth cannot seal out water. When water goes into the mouth, a valve at the back of the throat (just in front of where the nasal passages open from the secondary palate) seals off the throat (fig. 1.5). Thus, a combination of the nostrils closing, the throat valve closing, and the existence of the secondary palate keep the nasal passages and lungs free of water when submerged.

This arrangement also allows the animal to breathe when holding prey in its mouth or when only the nostrils protrude above the water's surface. The closing of the ears, nasal passages, and throat flaps are involuntary movements, so the alligator does not have to consciously think about doing all of these things each time it goes underwater. Respiration itself (taking in oxygen and expelling carbon dioxide) is another example of an involuntary movement or reflex. When the alligator is diving underwater, blood flow is reduced except for the oxygenated blood that goes to the heart and brain.

Alligators are propelled as they move through the water by their laterally flattened tail that makes S-shaped undulations through the water. The rounded snout parts the water (fig 1.6), and the torpedo-shaped body reduces drag, as do the four limbs being held against the body. Although the top of the body is heavily armored with bony plates for protection, the

Figure 1.3. The "nictitating membrane," a third eyelid, goes across one side of the eye to the other like a curtain when the alligator is underwater. It serves to waterproof and protect the eye.

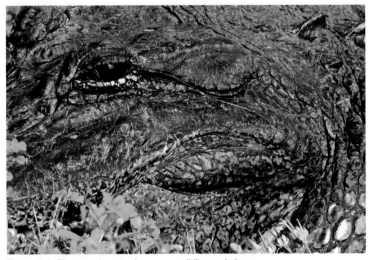

Figure 1.4. The openings to the ear are slitlike and close when the alligator submerges.

a

Figure 1.5. (a) The mouth open above water. Note the horizontal boundary of the tissue in back of the tongue that will seal when the alligator is underwater. (b) Throat valve in a sealed position as it is when the animal is underwater.

b

Figure 1.6. The
alligator's rounded
snout parts the
water and, along
with its torpedo-
shaped body,
reduces drag.

sides of the body and the belly are smooth and
also reduce drag (fig. 1.7). The color, shape,
and the bony armor give the alligator a loglike
appearance at times, affording it camouflage in
its habitat (fig. 1.8).

When not actively moving in water, the
alligator can retain buoyancy by manipulat-
ing lung volume and respiration. One position
is for the top of the entire body to be horizontal at (i.e., parallel to) the water's surface with
the limbs stretched out to the sides, serving
as stabilizers. Sometimes the whole dorsum
(top) of the body is visible; at other times the
head and high point of the back (and perhaps
the end of the tail) are the only portions of the
body visible. A second position is for the head
to be at the water's surface with the rest of the
body slanted down into the water somewhat
vertically so that the tip of the tail is the closest
body part to the substrate or bottom. The first
position is advantageous in that the alligator is ready to resume swimming. The second
position allows the alligator to quickly "back
down" into the water, for example, if it gets
"spooked" by something. Because the nostrils
are the last portion of the top of the head to
go underwater, the alligator has time to take a
final breath before submerging.

When the alligator moves on land, it uti-
lizes a gait known as the "high walk" in which

Figure 1.7. The torpedo-shaped body and the smooth sides and belly help reduce drag when the alligator is swimming.

the head, all four legs, and sometimes even the tail are carried well off the ground (fig. 1.9). The posture near water is more like that of lizards with legs lowered and belly often dragging. Feet that possess claws are associated with land animals, yet in the case of the alligators and other crocodilians, claws are very useful at the land/water interface where they help the animal maintain footing on slippery banks. All four feet are partially webbed, and this also is advantageous in the amphibious environment, where it aids in walking in shallow water and on muddy banks (fig. 1.10).

The alligator has many other adaptations to its habitat, especially in utilizing its environment to thermoregulate by a combination of behavioral, physiological, and morphological means. Thermoregulation and other adaptations that relate to the life history are discussed in chapter 3.

Crocodilian Diversity of the Past

From fossilized remains, we can ascertain information regarding species diversity and distribution of extinct crocodilians. The first crocodilians appeared in the fossil record at what is believed to be about 215 million years ago in the Mesozoic era at the interface of the Triassic and Jurassic periods, when dinosaurs and other reptiles were most populous (see table 1.3). Birds and mammals, as well as advanced bony fishes, most living insects, and flowering plants, first appeared in this era. The earliest

Figure 1.8. The alligator's loglike appearance is an adaptation that allows it to be camouflaged in its environment.

crocodilians were only about 3 feet (0.9 m) in length, with short snouts and relatively long limbs. They were lizardlike in appearance with an armored body and were thought to be terrestrial carnivores. Some species may have been semi-aquatic, but the group as a whole was not suited for an aquatic existence.

About 190 million years ago in the Lower Jurassic, crocodilian diversity increased. Sizes ranged from small species to those over 30 feet (9.1 m) in length. Species occupied terrestrial, freshwater, and marine habitats. Some marine forms were particularly well represented in the fossil record, including crocodilians that lost their armored body, developed paddlelike limbs, and had a small dorsal fin near the tail that was shaped like that of a shark.

The present crocodilian line, including the crocodile and alligator subfamilies, is known from the Upper Cretaceous in western North America approximately 80 million years ago. The genus *Alligator* is known from the Miocene of Florida; thus, Florida is believed to be the center of origin of the American alligator.

Figure 1.10. Claws and partial webbing on the feet are advantageous at the land-water interface.

The Chinese alligator appeared in the Pleistocene of China, but its ancestral relationship with the American alligator is unknown.

Crocodilian Distribution in the Past

The study of the past and present distribution of species in the world is known as biogeography. Two factors are considered to be very important in crocodilian biogeography. First, all crocodilian species, living or extinct, terrestrial, freshwater, or marine, needed a warm climate in which to live. Second, continental drift, the coming together and separating of the continental landmasses through time, played a role. If one looks at a world map, it is possible to see how the present continents could fit together, especially Africa and South America. The earth's crust and upper mantle are made up of large plates that move by convection currents created by the molten mantle inside the earth (plate tectonic theory). If two plates separate and there is a landmass that is on both plates, then that piece of land will be split. Geologic

Table 1.3. Geologic timescale

*LOCATION OF CONTINENTS	ERA	PERIOD		EPOCH	**AGE (MILLIONS OF YEARS AGO)	KEY EVENTS
	CENOZOIC	Quaternary		Holocene	0.01	Historical time
				Pleistocene	2.6	Ice ages; humans appear
		Tertiary	Neogene	Pliocene	5.3	Bipedal human ancestors appear
				Miocene	23.7	Continued radiation of mammals and angiosperms; earliest direct human ancestors appear
			Paleogene	Oligocene	33.9	Many primate groups appear including apes
				Eocene	56	Angiosperms widespread; continued radiation of modern mammalian orders
				Paleocene	66	Major radiation of mammals, birds, and pollinating insects
	MESOZOIC	Cretaceous			145	Flowering plants (angiosperms) appear; mass extinction at end of Cretaceous includes dinosaurs
		Jurassic			201	Gymnosperms are the dominant plants; many types of dinosaurs are abundant
		Triassic			252	Cone-bearing plants (gymnosperms) are dominant; many new types of dinosaurs appear
	PALEOZOIC	Permian			299	Mass extinction of marine and terrestrial life; radiation of reptiles and most modern insect orders
		Carboniferous			359	Origin of reptiles and seed plants; amphibians and vascular plants dominant
		Devonian			419	Origin of amphibians and insects; radiation of bony fishes
		Silurian			444	Origin of jawed fishes and diversification of jawless fishes; diversification of early vascular plants
		Ordovician			485	Marine algae plentiful; land being colonized by plants and arthropods
		Cambrian			541	Cambrian explosion; diversification of many phyla of modern animals
	PRECAMBRIAN				635	Many kinds of soft-bodied invertebrate animals; many types of algae
					2100	Oldest fossils of eukaryote cells
					2,700	Atmospheric oxygen starts accumulating
					3,500	Oldest cell fossils found (prokaryotes)
					3,800	Oldest known rocks on Earth's surface
					4,600	Approximate origin of Earth

*The continents drifted apart over time, resulting in changes in the climate and the distribution of living organisms. (Illustrations by Lisa McDonald)

**Ages given start from the beginning of periods and epochs.

events such as mountain building, the creation of oceanic islands and oceanic basins, volcanic eruptions, and earthquakes all take place at plate boundaries. So-called faults, such as the San Andreas Fault in California, are at the interface of two sliding plates.

The continents had come together an estimated 200 to 250 million years ago at the interface of the Paleozoic and Mesozoic eras to form one huge continent known as Pangaea. Thus, the first crocodilians were known at a time when there was one continent, which explains why the early crocodilians were widespread. About 135 million years ago during the Cretaceous, Pangaea began to split into two great landmasses, Laurasia to the north and Gondwana to the south. From fossils dated approximately 115 million years ago, it is known that both continents contained similar species of terrestrial and freshwater crocodilians. By an estimated 100 million years ago, a divergence of crocodilian species was seen in all but the marine forms on the separate landmasses. When the alligator and crocodile lines became known some 80 million years ago, the major continents of today were formed to the extent that none of the alligator subfamily was present in Africa. At present, only two alligator species remain.

The American Alligator: Living Fossil?

One of the perceptions that people commonly have about the alligator and other living crocodilians is that they are survivors from a prehistoric age, just hanging on. Fern plants are a much older group (from the Devonian period of the Paleozoic era) than the crocodilians, yet it is not common to look at a fern plant and marvel how it preexisted and coexisted relative to the dinosaurs. There are about 12,000 species of ferns and only some 20 species of crocodilians. Furthermore, crocodilians have a limited range on the earth, whereas ferns have a much greater distribution. Crocodilians are amphibious and may live in large expanses of water away from most civilization, thus restricting most people's exposure to them. Also, many people fear reptiles and typically stay away from them. And a big, impressive carnivorous reptile incites a number of strong emotions, including fear and awe, in the general populace. Most humans in the world do not coexist with crocodilians in the sense that someone does in, say, Port Arthur, Texas, who lives next to a marsh and makes a living by hunting and fishing in that marsh. That lack of familiarity itself makes crocodilians an oddity.

The "surviving fossil" idea is further fueled by the fact that crocodilians are the only living reptiles to belong to the subclass Archosauria, which includes the impressive (but extinct) dinosaurs. Another reason that alligators and other crocodilians are called "living fossils" arises from their diverse past and relative lack of diversity at present. At one time there were terrestrial, marine, and freshwater species and many more species than exist today. After the crocodilians' habitat was no longer strictly terrestrial, they became primarily freshwater animals with the exception of some terrestrial and marine forms. When the dinosaurs became extinct at the end of the Cretaceous period, the crocodilians as a group were relatively unaffected and were in fact prospering, possibly because whatever caused great extinctions on land and in the seas did not affect the food chain in freshwater habitats. Also, crocodilians can go without eating for extended periods of time and can "wait out" many adverse conditions. Finally, it is important to realize that the crocodilian species alive today are not the same ones that lived with the dinosaurs. And above all, the American alligator and other crocodilian species of today are superbly suited for their habitat—where land and water meet.

2 Where Do Alligators Live?

With very few exceptions, the meeting place of land and water has always been home for the crocodilians. Drenched with sunlight and richly supplied with plant and animal life drawn from both water and land, these areas are cornucopias for the large amphibious predator. The footing is too uncertain for the big meat-eaters of the land, while the big fish are in danger of stranding or becoming trapped in pools that become fatally depleted of oxygen. Independent of the water when he has to be but able to cruise it for prey or escape into it from enemies, the crocodilian makes the best of two worlds.

—Sherman Minton and Madge Minton, *Giant Reptiles*

Alligators occupy a variety of natural or human-made water bodies and are tolerant of an array of habitat conditions (fig. 2.1). As a result, they have greater flexibility and success than many other species that cannot adapt as well. Alligator distribution is very confusing because the population declined greatly and later rebounded. It is hard to discern where alligators have always been, where they previously have been but are now absent, and what is new territory that they have invaded. Undoubtedly, there are counties that historically contained healthy alligator populations that were known by the locals but were never reported in the scientific literature. For example, Fort Bend County, where Brazos Bend State Park is located, was absent from Raun and Gelbach's 1972 map. I have attempted to update all known county records (table 2.1 and fig. 2.2).

Historical versus Present Distribution in Texas

The 1984 Texas Parks and Wildlife Department's (TPWD) Management Plan for the American Alligator in Texas stated that alligators currently occurred in more than 90 % of the historic range described by Raun and Gehlbach. It listed 53 additional counties where alligators had not been reported by Raun and Gehlbach. TPWD noted that the greatest concentrations of alligators were found in the middle and upper coastal counties, but significant populations existed inland in suitable hab-

itat. The distribution "encompasses all major river drainages in the Gulf Coastal Plain, their tributaries, wetlands, lakes, stock tanks, and farm ponds" (Thompson, Potter, and Brownlee 1984).

Mittleman and Brown (1948) predicted the extinction of the alligator. They were trying to define the historical range of the alligator in Texas by both boundaries and counties before the predicted extinction actually occurred. Additionally, they had reports of counties with reliable records of the existence of one or more animals. Mittleman and Brown quoted Stejneger and Barbour (1943), who gave the western limit of the alligator's range as the Rio Grande in Texas. Mittleman and Brown stated that older records confirmed this, which is both the US and Texas range limit, but pointed out that the alligator did not occur in the Rio Grande at the time of their writing in 1948. Mittleman and Brown determined the westernmost limit of the alligator to be in Kinney County (following Mearns 1907). They noted that Strecker (1915) believed the whole eastern half of Texas to be the original range of the alligator, but that at the time of his writing the

Figure 2.1. Crocodilians live where land and water meet.

Table 2.1. Counties in Texas with records of alligators

Anderson	Gregg	Nacogdoches
Angelina	Grimes	Navarro
Aransas	Hamilton	Newton
Atascosa	Hardin	Nueces
Austin	Harris	Orange
Bastrop	Harrison	Panola
Bee	Hays	Polk
Bexar	Henderson	Presidio
Bowie	Hidalgo	Refugio
Brazoria	Hopkins	Robertson
Brazos	Houston	Rusk
Brewster	Hudspeth	Sabine
Brooks	Jackson	San Augustine
Brown	Jasper	San Jacinto
Burleson	Jefferson	San Patricio
Burnet	Jim Wells	Shelby
Calhoun	Karnes	Smith
Cameron	Kaufman	Titus
Camp	Kimble	Travis
Cass	Kinney	Trinity
Chambers	La Salle	Tyler
Cherokee	Lamar	Upshur
Collin	Lavaca	Val Verde
Colorado	Lee	Van Zandt
Dallas	Leon	Victoria
DeWitt	Liberty	Walker
Delta	Limestone	Waller
Dimmit	Live Oak	Webb
Duval	Llano	Wharton
Ellis	Madison	Williamson
Falls	Marion	Wood
Fannin	Matagorda	Zapata
Fayette	Maverick	Zavala
Fort Bend	McLennan	
Franklin	McMullen	
Freestone	Medina	
Frio	Milam	
Galveston	Montgomery	
Goliad	Morris	

Note: Data for 111 counties from the literature and/or specimens in collections.

species was found chiefly in the eastern and southeastern counties bordering Louisiana and the Gulf of Mexico.

Records of alligators in the wild in Texas that Mittleman and Brown considered reliable through 1935 included the following Texas counties: Bexar, Bowie, Cameron, Dallas, Duval, Falls, Henderson, Jefferson, Jim Wells, Kinney, Leon, Liberty, Live Oak, McLennan, McMullen, Montgomery, Nueces, Refugio, Robertson, Travis, Victoria, and Williamson. They updated their list of the counties where alligators still existed at the time of their publication in 1948, and alligators were known to exist only in Dallas, Leon, and Victoria Counties. However, they knew that alligators existed in several other counties, which they added: Anderson, Atascosa, Colorado, Harris, Hamilton, and Hays. These updates were largely the result of reports from Texas Herpetological Society members.

Grant (1956) also listed Val Verde County for the range of alligators in Texas, which extended the western range by one county past Kinney, if it was accurate. Grant also noted the sighting of alligators along the Rio Grande in Maverick County, located directly south of Kinney County.

Raun and Gehlbach's (1972) compilation of distribution maps for Texas reptiles and amphibians included all literature records that they considered reliable as well as voucher specimens that they had examined in museum collections. They did not include any counties along the Rio Grande for the alligator and in the text account referred to these "records of former occurrence" as "questionable."

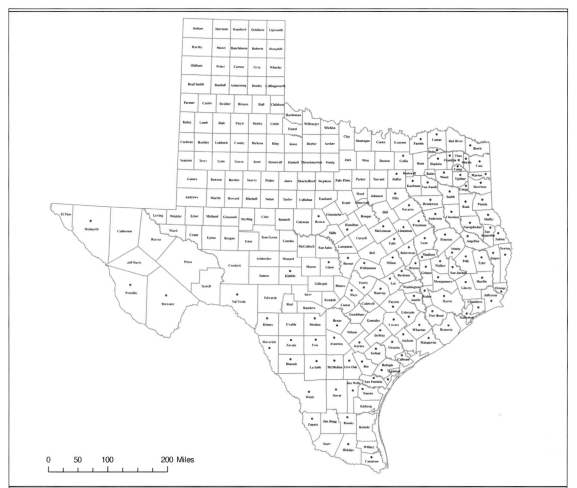

Figure 2.2. Range map of Texas showing the counties where alligators are known to occur.

Jean-Louis Berlandier, a French botanist who explored Mexico and Texas in the early 1800s, asserted that the alligator definitely did not occur in the Lower Rio Grande Valley and not south of the Nueces River. That would exclude those three counties along the Rio Grande. His 1,500-page book on his travels in Mexico and Texas from 1826 to 1834 was published in French and not translated into English until 1980. Smith and Chiszar (2003) published notes from this manuscript on the crocodilians, so Berlandier's opinion was being heard for the first time, albeit some 170 years later.

Another early work that has not previously been cited for Texas alligator distribution is Captain Flack's book published in 1866. He saw an alligator at some lakes about 6 miles (9.7 m) from the town of Columbia on the Brazos River, which is present-day East Columbia in Brazoria County. This would have added Brazoria County to the alligator's range long before Raun and Gehlbach's work. Another new county reference was for Galveston County, where Flack saw a large alligator in the city of Galveston that had been hauled into town from a nearby lake. Galveston County was not included in Raun and Gehlbach's work but was added to county records later on.

Factors Influencing Alligator Distribution

So why do alligators occur where they do? Factors named by Neill (1971) as influencing the entire US distribution of alligators are the sea, the fall line, temperature, and precipitation. Since alligators are freshwater animals (although they inhabit some saline habitats), the sea would serve as a limit to their range to the south and the east. The fall line is a line connecting the first major series of waterfalls or rapids that exist inland on any major waterway from the coastline. During the Cretaceous period, the sea covered the lowlands of the eastern United States. The fall line marks the boundary of the shoreline of the former seabed, dividing it into the Piedmont Plateau (the portion that did not possess water) and the Coastal Plain (the formerly submerged area). The Balcones Fault is the western boundary of the Texas Gulf Coastal Plain and is usually considered the boundary between the lowlands and uplands of Texas. It extends from a point near Del Rio eastward to northwestern Bexar County and then northeastward to the Red River. The alligator and many other animals

and plants that inhabit the Coastal Plain find their northern range boundary to be roughly at the fall line, which serves as the southern range boundary for many species that live in the Piedmont Plateau. Of course, a number of species inhabit both areas, such as the raccoon, because they are not as rigidly tied to a specific habitat (i.e., they tend to be generalists).

As mentioned in chapter 1, cold temperatures and crocodilians do not mix. A general temperature cutoff point for alligator distribution to the north is represented by an isotherm of 35°F (1.7°C) of mean January minimum temperatures. An isotherm is a line depicting geographic locations that are the same in some aspect of temperature and are usually drawn at intervals of 5°F (−15°C). The concept of microclimate (the actual weather conditions in a certain location or small area that are different from the climate of the surrounding area) comes into play and can allow for alligators to extend farther north in some specific localities. Alligators in the northernmost portions of their range may not be able to survive a colder than normal winter.

The lack of rainfall to the west and south in Texas is another factor that affects the alligator's range. However, as alligators can move via rivers that flow through arid regions, stragglers can show up in parts of Texas that would seem inhospitable. This probably accounts for some of the county records that consist of lone alligators. It is important to keep in mind that the presence of a wild alligator in such a location is certainly not indicative of a breeding population but rather of the amazing distances that alligators are capable of traveling. Other sightings of alligators in unexpected locations could be due to deliberate or accidental releases of captive alligators.

Types of Water Bodies That Alligators Inhabit

Several schemes can be used for describing alligator habitat types. Popular field guides tend to note the array of water bodies where alligators may be encountered within their range. Conant and Collins (1991, 39) list "great river swamps, lakes, bayous, marshes, and other bodies of water" as the places where they are found. Behler and King (1979, 429) name "fresh and brackish marshes, rivers, swamps, bayous, and big spring runs" as habitat. A marsh is defined as an area of wet, low-lying land (fig. 2.3), and a swamp is defined as a marsh, so the use of the two terms as separate entities is a bit subjective. However, there is a tendency to call certain types of marshes "swamps" based on their dominant vegetation (usually perennial trees and shrubs), such as cypress swamps. Note that Behler and King differentiate fresh and brackish marshes (containing salt) from each other.

A bayou is a marshy, sluggish body of water that is a tributary (a body of water that flows into another) to a river or lake. A stream is a body of water that runs in a channel, and a river is a large stream. Creeks and sloughs are

Figure 2.3. Pilant Lake at Brazos Bend State Park, an inland marsh.

other habitats of alligators. A creek is a small stream (fig. 2.4), and a slough is a slow-moving, marshy backwater (fig. 2.5).

Lakes are large bodies of water surrounded by land, and a pond is a body of water surrounded by land that is much smaller in size. An oxbow is a special type of lake formed when a river changes its course and leaves behind crescent- or horseshoe-shaped channels of water (fig. 2.6). A big spring run refers to the water flowing from a big spring and is a more likely habitat type for Florida (e.g., Silver Springs) than Texas.

Figure 2.4. Big Creek at Brazos Bend State Park. This creek is at a full water level and can shrink in size with reduced rainfall.

Figure 2.5. Pilant Slough at Brazos Bend State Park. During drought this backwater habitat may dry up in many sections and drive its inhabitants to permanent water bodies.

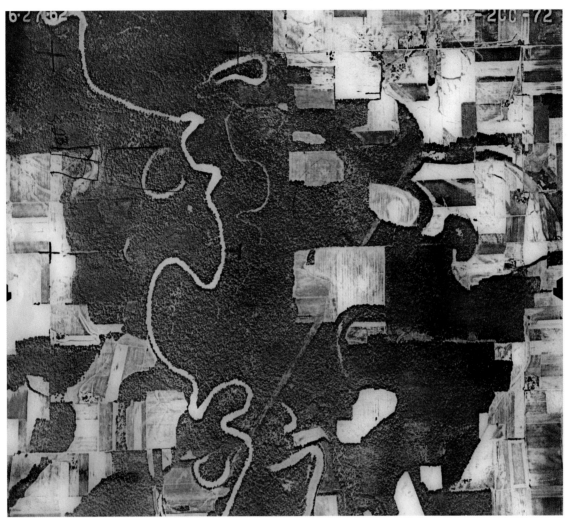

Figure 2.6. A river changes its course and leaves behind crescent- or horseshoe-shaped channels of water known as oxbows. (Aerial imagery courtesy of the United States Department of Agriculture Farm Service Agency)

Coastal versus Inland Habitat

TPWD, the State of Texas regulatory agency for alligators, splits alligator habitat within the state into "coastal" and "inland." Coastal habitat consists of coastal marshes (fig. 2.7), which are further subdivided into saltwater marshes, brackish marshes, fresh to brackish marshes, and freshwater marshes. Inland habitat is sub-divided into freshwater rivers (rivers, creeks, and ditches), swamps, and marshes (marshes and bayous); and natural and human-made ponds and lakes (lakes, ponds, and reservoirs).

Prime versus Marginal Habitat

Prime habitat consists of high-quality habi-tat for alligators that meets their basic needs

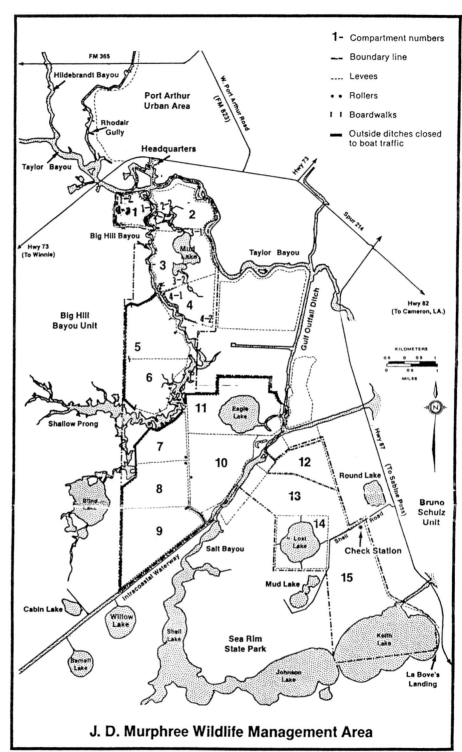

Figure 2.7. (a) Map and (b) aerial photo of prime coastal marsh at the J.D. Murphree Wildlife Management Area. (Aerial imagery courtesy of the National High Altitude Photography program)

b

for food, water, and shelter and is configured in tracts large enough to support a substantial reproducing alligator population. Large areas of coastal habitat typify this definition, but inland habitat can be spottier. An inland habitat that is prime needs to consist of either an extremely large body of water (i.e., a major lake) or a composite of inland water bodies such as marshes, lakes, ponds, bayous, streams, and creeks associated with a major river so that migration in and out of the site is possible. TPWD noted in its 1984 report that the presence of such dispersal routes in inland habitat tends to maintain healthy, stable alligator distribution throughout available suitable habitat.

The J. D. Murphree Wildlife Management Area is an example of prime coastal habitat (see fig. 2.7). It consists of 12,386 acres (5,012.5 hectares) of fresh and brackish marsh located in Jefferson County near Louisiana. It is divided into 11 compartments through a levee system north of the Intracoastal Waterway, and south of the waterway it consists of 2 compartments without levees that are subject to tidal fluctuations. The northern compartments make up a vast expanse of coastal habitat where a sizable population of alligators can live and reproduce.

Brazos Bend State Park is prime inland habitat of the composite type (fig. 2.8). The park, formerly known as Hale Ranch, consists of 4,687 acres (1,982 hectares) in Fort Bend County. The prominent water bodies in the park include a freshwater marsh, three lakes,

three oxbows, a slough, and a creek. The Brazos River is present at the eastern boundary of the park. Most of the park is in the Brazos River floodplain, but there are also areas of forest above the floodplain and tallgrass prairie.

The alligator population is at or near the carrying capacity of the park (i.e., the largest number of alligators that the resources of the park can support indefinitely). Consequently, some alligators at Brazos Bend live in areas of the park that might be termed marginal habitat, such as prairie potholes and intermittent streams. The prairie pothole area at Brazos Bend (fig. 2.9a, b) has supported two nesting females and associated juveniles, and a nearby detention pond has one nesting female and associated juveniles. The prairie pothole area is subjected on a nearly annual basis to extremes in water levels that are too high or too low. Alligator holes with associated dens (fig. 2.9c) hold the last remnants of water when the potholes dry up, but sometimes even the holes and

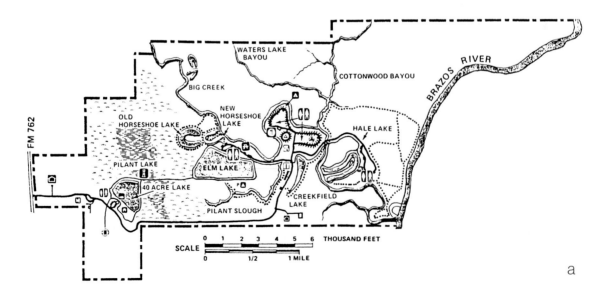

a

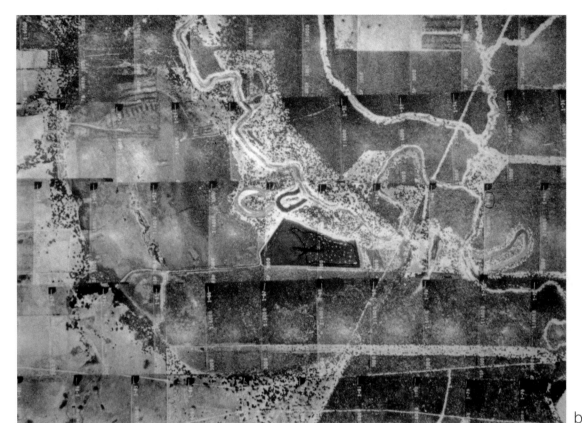

b

Figure 2.8. (a) Map and (b) aerial photo of prime inland habitat at Brazos Bend State Park with a variety of types of water bodies. (Aerial imagery of Hale Ranch State Park, which is now known as Brazos Bend; photographs and support documentation: Texas Parks and Wildlife Department Infrastructure Division Records, Archives and Information Services Division, Texas State Library and Archives Commission)

the den opening do not contain any water (fig. 2.10).

The most adverse effect of drought and flooding at the pothole area is on nesting. Nests can get flooded out and the embryos drown, and when there is no water, both embryo and hatchling mortality is high. Footprints indicate that adult alligators may leave the pothole area during dry spells, presumably to hunt for food and get to water. When water is present, mammals such as deer and raccoons, snakes, crayfish, insects, and insect larvae appear to be plentiful and provide an adequate food supply for all sizes of alligators. Intermittent streams may be filled with water or even flooded at times. During a portion of the year they tend to have only a trickle of water running through them (fig. 2.11). At Brazos Bend, at least one nesting female is known to be associated with an intermittent stream. It would seem that at times of low water food might be in short supply for a group of young alligators that would

have to compete among themselves for small prey living in or near the stream.

Marginal habitat can be located at the boundaries of the alligator's range, where factors such as low winter temperatures might kill individuals who may have been able to survive through a couple of mild winters previously. It can be an estuary having salt concentrations that are too high for alligators to live in but may serve as a place to hunt for prey. Some marginal areas may contain suitable habitat but be too small in size to support more than one or two resident alligators. Marginal habitat may be used seasonally or for a dispersion route.

Use of Human-Made Habitat

The use of human-made habitat is variable, as is its quality. Reservoirs are large, human-made lakes where water is held for a community water supply or for irrigation. They can support permanent, successful populations of alligators if shallow areas are available for nesting and rearing juvenile alligators and sufficient food is present. Tanks or farm ponds are dug in rural areas to water cattle, raise game fish, or serve as decoration. They may have the habitat attributes to provide food, shelter, and nesting areas, but size limitations allow them to support no more than a couple of adult alligators over time. There are also a number of detention ponds being built in suburban areas so that the "concreting" of the area does not result in flooding. These ponds may eventually play host

Figure 2.9. Brazos Bend State Park. (a) The prairie pothole area. (b) A pothole. (c) An alligator hole at the prairie pothole region.

Figure 2.10.
A dry den site
at the prairie
pothole area at
Brazos Bend
State Park.
Alligator holes
with associated
dens are the last
places to retain
water during
drought.

to alligators, at least as stopping sites during migration. Drainage ditches, drainage pipes, and water-filled depressions can hold alligators, especially seasonally, of all size classes (fig. 2.12).

Drainage pipes act as alligator dens of sorts and can be used as refugia from heat and cold both with and without water (fig. 2.13). Rice canals have the potential to be a permanent home if prey can be found on a regular basis. Swimming pools are not really habitat but can hold alligators temporarily while migrating to other areas. Neglected pools may even have large insect larvae, adult insects, and resident adult bullfrogs that can provide a food supply.

Wetlands and Deepwater Habitats

Even formal systems to classify types of water bodies have been fraught with problems and conflicts. The US Fish and Wildlife Service (USFWS) published *Classification of Wetlands and Deepwater Habitats of the United States*, a classification system developed by wetland ecologists (Cowardin et al. 1979). This serves as a system that can be utilized by geologists, limnologists, botanists, and wildlife biologists to make terms and concepts more uniform and thus allow statistically tenable calculations to be more easily made (i.e., get everyone on the same page so that comparison of sites and

Figure 2.11.
An intermittent
stream at Brazos
Bend State
Park at a time
when it contains
very little water.
Intermittent
streams can
be full of water
and have a
substantial
current or be
reduced to a
trickle when
water levels are
low.

Figure 2.12. This
juvenile alligator
was found in
this water-filled
depression near
the entrance of
Brazos Bend
State Park.
These low-lying
areas hold water
after heavy
rains and are
frequented by
juveniles that are
dispersing.

trends can be determined with some accuracy). The system is a hierarchy, a graded series from system, class, and subclass, to identifying dominance types in terms of plant and animal species. Unlike the field guides, it differentiates between wetlands and deepwater habitat. It defines wetlands as "lands transitional between terrestrial and aquatic systems where the water table is usually at or near the surface or the land is covered by shallow water." Deepwater habitats are "permanently flooded lands lying below the deepwater boundary of wetlands."

The five systems are marine, estuarine, riverine, lacustrine, and palustrine.

The marine system is not considered to be a habitat for the American alligator. Some habitat near the coast falls into the category of the estuarine system, but the bulk of alligator habitat in Texas belongs to the riverine, lacustrine, and palustrine systems.

We can use a portion of Elm Lake at Brazos Bend State Park that has deep, relatively open water as an example of how the system works (fig. 2.14). This is the classification from the

Figure 2.13. Drainage pipes with or without water serve as shelter for alligators. (Photo from the author's collection)

category of "system" through the category of "dominance type."

> System: Lacustrine
>
> Class: Aquatic Bed
>
> Subclass: Rooted Vascular (referring to a plant that is rooted in the bottom and has a conducting system that serves to transport fluids to various parts of the plant)
>
> Dominance Type: Water Lily/Lotus

Figure 2.14. This portion of a lake at Brazos Bend State Park exhibits a plant dominance type of water lily/lotus.

Life History of Alligators

The behaviour of crocodilians is an enigma. To the casual observer, these giant lethargic reptiles are the antithesis of "behaving" animals because, most of the time they exhibit little or no activity. Indeed, it is their immobility in the face of persistent attempts to arouse them which usually impresses onlookers. To those who live or work around alligators or crocodiles, the most notable behaviours are likely to be the sudden movements associated with feeding and with threat. Displays of strength and cunning are infrequent and unpredictable—but when seen, they are unforgettable and a constant reminder that there is more to the behaviour of crocodilians than meets the eye. In fact, crocodilian behaviour is subtle, complex and quite unlike that of other reptiles.

—Jeffrey W. Lang, *"Crocodilian Behaviour: Implications for Management"*

Why does that alligator just lie around all day doing nothing? What does an alligator do during a flood or hurricane? How long can an alligator stay underwater? Why does that alligator have a short tail/is missing a limb/has a lump on its nose? How fast does it grow? How big does it get? What does an alligator eat? Does an alligator get sick? What are the causes of mortality or death? How long can an alligator live? These are questions that I am commonly asked about alligators. Scientists ask some of these same questions in the disciplines of ethology (animal behavior), environmental physiology (the functions and processes that occur inside an animal's body as they relate to its external environment), and wildlife biology relating to alligators.

This section is the longest in the book and combines research findings with the answers to questions that the general public has about the only Texas crocodilian. Many of these research findings are the result of my own scientific endeavors with Texas alligators, primarily at Brazos Bend State Park near Houston and at the J. D. Murphree Wildlife Management Area (WMA) at Port Arthur. Many of the natural history observations that I have made are findings not conducive to publication in a scientific journal. These observations provide insight into the life of the American alligator, and this book is a suitable vehicle for sharing this knowledge with both scientists and non-scientists.

Basking/Thermoregulation

The fallacy that crocodilians are lazy has been perpetuated through time. Reptiles are poikilo-thermic—they lack the capacity to maintain a stable body temperature, so their temperature varies with that of the environment, which makes them ectotherms. The commonly used nonscientific term is "cold-blooded," which is misleading, as this suggests that reptiles are always colder than endotherms ("warm-blooded" animals such as birds and mammals that are able to internally regulate a stable body temperature).

Poikilotherms can be divided into two types based on how their body temperature is regulated. Fish are passive thermoregulators because they "make do" with the tempera-ture of the water that they live in and thus are adapted for the temperature range of their par-ticular habitat. Although they produce some internal metabolic heat via their body pro-cesses, the elevation of body temperature over water temperature is usually slight because water absorbs heat extremely well and fish are poorly insulated.

Crocodilians and most other reptiles are active thermoregulators, which means that they regulate their body temperature from being too hot or too cold by taking in heat from the envi-ronment or giving it off to the environment. In order to regulate their temperature, they behav-iorally seek a certain range of temperatures.

Since alligators and other crocodilians live both in water and on land, this behavioral ther-moregulation involves going back and forth between the two habitats. For example, in the cool seasons of the fall, late winter, or early spring when the air temperature is warmer than the water temperature on sunny days, the alligator leaves the water to bask in the sun (fig. 3.1).

When it becomes cooler as the sun goes down, the alligator returns to the water. Con-versely, in the summer, alligators stay in the water during the heat of the day, as it is cooler than being out on land even in the shade, and then leave the water in the evening and early morning when the air temperature is lower. Alligators also remain in the water during the winter when the air temperature is too cold to provide any warmth. When alligators take ref-uge in the water on hot summer days or in the winter, they remain submerged or have only their head showing. In the summer, the upper layer of the water heats up, while the lower layer of water and the muddy substrate are cooler (fig. 3.2). In the winter, the lower layer of water and the substrate are warmer than the upper layer of water. Water heats and cools more slowly than land so helps buffer tempera-ture extremes. However, extremely shallow water can get very cold or hot. Consequently, the alligator needs a water area about 3 feet (0.9 m) deep for the water to serve as an effec-tive buffer.

Both living organisms and nonliving objects use four physical mechanisms to gain

heat from or lose it to the environment: radiation, conduction, convection, and evaporative cooling. Radiation involves the transfer of heat without direct contact with a heat source. Crocodilians can warm up through the radiant heat of the sun when they bask or lose heat to cooler air surrounding them. Conduction involves the transfer of heat from a warmer object to a cooler object that is in direct contact with it. In crocodilians this direct transfer occurs when the reptile is too hot, goes into the water, loses heat to the water, and its body temperature

cools down as a result. The crocodilian can also heat up by making contact with a warm substrate (called thigmothermism). Alligators are known to move during basking, which results in their underside making contact with a surface warmed by the sun. This can speed up heating since the radiant heat of the sun cannot reach beneath the body. Also, when they are basking and the sun begins to go down, if there is a substrate that is still warm, they will gravitate toward it. Thus, alligators are attracted to certain human-made structures such as roads,

Figure 3.1. Alligators are behavioral thermoregulators and move between water and land to regulate their body temperature. The typical basking season in Texas is in the fall, late winter, and early spring.

Figure 3.2.
The muddy
substrates of
water bodies
help insulate
alligators
from both
hot and cold
temperatures.

parking lots, and walkways that retain heat after the air temperature cools.

Convection is the movement of air or water in the form of currents. Air blowing around an animal causes the warmer air to rise and expand; the denser, cool air (nearer the alligator) absorbs heat faster than air that is already warmed. The result is that the animal cools down. Evaporation occurs when water molecules absorb heat from the environment and become energetic enough to form a gas that escapes into the air, taking a large amount of heat with it. Crocodilians do not have sweat-producing glands, but evaporative cooling can take place when their wet backs are

exposed to the warmer air or they open their mouths (mouth gaping) when basking and the moisture inside evaporates and cools them off (fig. 3.3).

In captivity, reptiles should be given a thermal gradient so that they can behaviorally thermoregulate much as they would in the wild. Usually, one end of the cage or enclosure is at the lower end of the preferred temperature range, and a spot at the other end is at the upper end of the preferred range, a basking spot under a light or a substrate with a heated rock or heat pad. The reptile can choose the optimum temperature for its particular situation at the time by moving between the temperature

Figure 3.3.
Mouth gaping
while basking
can help in
evaporative
cooling as
the moisture
inside alligators'
mouths
evaporates and
helps cool them
off.

gradients. For example, after a large meal, a reptile typically seeks out additional heat to help digest its food. An inactive reptile would tend to be in the cooler portion. An outdoor enclosure should be designed so that there is a sunny portion and a shady portion with areas having a range of temperatures.

Size Differences and Physiological Thermoregulation

A large alligator has more insulation than a small alligator, as well as less surface area relative to its total body size. The smaller an animal is, the larger its surface-to-volume ratio. The transfer of heat is proportional to the surface area exposed. As a result, a small alligator or other small poikilotherm passively heats and cools much more rapidly than an adult alligator or other large poikilotherm and is also largely subject to the effects of conduction via wind (fig. 3.4).

One might expect that the body temperatures of small and large alligators would differ. Such is not the case. Newly hatched crocodilians are known to maintain a higher body temperature (presumably to help them absorb their residual yolk), but there does not appear to be a difference in body temperatures of juveniles and adults when measured under the same climatic conditions. This temperature

Territorial Basking Counts

Brazos Bend State Park is the ideal location to test using a census method during the daylight hours at the height of the season when alligators are most likely to be basking for prolonged periods. From 1985 to 1989, I conducted several basking counts per year to monitor the number and sizes of alligators viewed over several years, to compare these day counts with TPWD night counts, to examine monthly differences in alligator visibility and location (land versus water), to pinpoint environmental factors that would best predict when the greatest number of alligators could be seen, and to make natural history observations (for the method to conduct a count, see Thompson and Gidden 1972).

I chose a route where the greatest number of alligators could be seen on foot by walking along footpaths between water bodies. Consecutive reference numbers were assigned to locations along the trail, usually corresponding to a structure such as a fishing pier, bridge, building, water control structure, or basking island. Census data included the date, time, air temperature, water temperature, wind (windy, breezy, or calm), percentage of cloud cover (cloudy, partly cloudy, partly sunny, sunny), numerical location, total length of alligator, position of alligator (in the water, on a log, on the bank), portion of the alligator in/out of the water, and comments.

The general trend was for alligators to bask the least from January to February and May to July, and to bask most commonly from March to April, with a peak in April. Alligators were usually in the water when sighted in January (nearly 100%) and July (approximately 70%), and typically only their heads were visible. From February through May, the entire body tended to be out of the water.

The percentage of adults and subadults (greater than 4 feet [1.2 m] in length) most likely to bask in February to May was almost twice that of juveniles. In January and July, the percentage of juveniles surpassed that of adults, and the difference was greatest in July. On cooler days when juveniles were observed, they were often found in dense stands of algae or aquatic plants, where temperatures were higher than that of the surrounding water.

Several statistical tests indicated that air temperature was the primary environmental factor that determined if alligators would bask. The minimum air temperature when alligators were observed to bask was 57.9°F (14.4°C), and the maximum was 84.9°F (29.4°C). The mean temperature (average) at which alligators basked was 78.3°F (25.7°C). Efforts to pinpoint conditions under which specific size classes were most visible were unsuccessful.

In evaluating these diurnal counts with state nocturnal data, which relies on seeing the alligators' red eyeshine, several points emerged. Since TPWD counts involved airboats patrolling both the interior and perimeters of the water bodies, it might be expected that these nocturnal counts would always be higher than day counts made by patrolling the perimeter on foot. The results indicated that for the larger water bodies, the perimeter counts were always much lower than the nocturnal counts by airboat. However, for a smaller lake, such as 40-Acre, diurnal perimeter counts could equal or exceed nocturnal counts with an airboat. Moreover, the sizes of alligators can be more accurately determined by day counts than by those conducted at night. Alligator population numbers by lake tended to have the same upward or downward trends in a comparison among years for both census techniques.

Obvious advantages to successful diurnal counts on foot are that night work is not required and considerable money is saved on airboat maintenance and fuel. Diurnal counts also can yield valuable natural history or behavioral data. During this study, I made many interesting observations of this type, including adults and juveniles feeding on various food items, feeding behaviors and strategies, cannibalism, breeding and other social behaviors between alligators, nest guarding and defense, adult interactions with pods (groups) of young, and interactions within pods. Alligators that were repeatedly associated with a certain location could be observed over a period of time, especially if there was an identifiable physical trait that distinguished them from other individuals.

The suitability of diurnal counts obviously varies among sites. This method would not be appropriate at coastal sites or other extremely large aquatic areas. A diurnal count was attempted at the Murphree WMA via airboat, since it was impractical to census the perimeter of a large coastal marsh on foot. Although the airboat allowed access to the vast marsh interior, the noise of the airboat scared the alligators, and few were sighted.

equity results from large and small alligators having somewhat different strategies to keep their body temperature in a preferred range (or a range as tolerable as possible when temperatures are extreme).

Large alligators can also thermoregulate by physiological adjustments to their blood flow and heart rate. Dilation of blood vessels increases blood flow, and constriction of blood vessels restricts blood flow. If a basking alligator wants to warm up, an increased blood flow to the superficial blood vessels allows the heat from the surface of the animal to be more quickly transferred to the core of the body. If an animal wants to conserve heat, then it would minimize blood flow to the outer part of the body.

Cardiovascular changes, such as an increase in heart rate (tachycardia) during heating and a decrease in heart rate (bradycardia) during cooling, also affect peripheral blood flow and are involved in thermoregulation. Mouth gaping exposes the interior of the mouth to the environment and may affect the rate of heat exchange between the head and the rest of the body. Regional heating and cooling can occur through increased or decreased circulation to certain parts of the body (i.e., increased circulation to the digestive organs after feeding). On the exterior of the alligator, the osteoderms (bony plates that make up the dorsal armor) have been found to possess a rich blood supply, which points to a role in thermoregulation as well.

Figure 3.4. It is not uncommon to see juvenile alligators pile up while basking. Since they heat and cool more rapidly than large alligators, this may be a behavioral strategy to insulate themselves by forming a larger mass as a group.

Preferred Temperature Ranges of Tropical and Temperate Crocodilian Species

Whereas the temperate alligator species encounter a range of temperature extremes from freezing winters to hot summers, the rest of the crocodilian species live in the warm tropics and primarily are concerned with avoiding heat and cooling off when possible. All crocodilian species use water for thermoregulation, but the patterns are different. Typically a tropical crocodilian will surface in the water in early morning, stay submerged during the heat of the day, and then come out of the water at night when the air is cooler. A typical regime for an alligator in the absence of extreme heat and cold would be to "prebask" in the shallow water in the early morning, spend the morning basking on land, return to the water in the afternoon with part of its back exposed to the air, and then remain in the water at night. This may be a response to photoperiod (day versus night within a 24-hour period) as well as temperature.

My personal observations of Cuban crocodiles and spectacled caimans in zoos is that they will not seek the warmer water in their outdoor enclosures at night when unseasonably cool temperatures occur during the summer in the north-central and central United States. They continue to leave the water at night, just as though it had been a warm day and they wanted to cool off. The significance of this behavior, as it relates to adverse climatic conditions in the wild, is that they would not be able to adjust to an extreme change in temperature and would likely perish. Tropical crocodilians in captivity should be able to be switched into winter quarters on a moment's notice. If that is not possible, the animals need to be chased back into the water and monitored to make sure that they do not leave it overnight.

How do these tendencies translate into actual recorded body temperatures? For alligators, the core body temperatures measured by E. H. Colbert and his team (Colbert, Cowles, and Bogert 1946) over a range of activities were 78.5°F to 98.5°F (26°C to 37°C). Lethal temperatures were between 100°F and 102°F (38°C and 39°C). Crocodilian biologist Jeffrey Lang (1987b) tested various species of crocodilians in thermal gradients and recorded the selection of mean body temperatures in the range of 77°F to 95°F (25°C to 35°C). For prolonged periods, temperatures below 59°F to 68°F (15°C to 20°C) were not tolerated.

Metabolic Rates and Exercise

Metabolism refers to all of the chemical processes that take place in the body in which energy is stored or released. The metabolic rate usually refers to the rate of energy consumption, or the rate at which the organism converts chemical energy to heat and external work. An alligator has approximately only 4% of a human of equal weight's metabolic rate. However, if the alligator heats up to 89.5°F (32.0°C), its metabolic rate will double. The smallest alligators with a body temperature of 82.5°F (28.0°C) have about half the metabolic rate of a human, and the largest ones have less than 2%. So there is an increase in metabolic rate with higher temperatures and a decrease with larger size.

Vertebrates utilize two types of metabolic activity: aerobic and anaerobic. Aerobic metabolism involves the continuous burning of fuel and oxygen consumption. Anaerobic metabolism does not require oxygen and can provide energy more quickly. However, the aerobic pathway is much more efficient at energy extraction, as it produces about 19 times more energy, and is utilized for prolonged physical activities. Crocodilians and other reptiles utilize anaerobic metabolism to move quickly and to undergo a lot of physical exertion for brief periods of time. Most of the time they tend to remain stationary or move very slowly, which, coupled with maintaining lower body temperatures when not feeding, minimizes the rate of aerobic metabolism and cuts oxygen use and fuel (food) consumption. The fact that crocodilians spend considerable time in the water, where they are buoyant, decreases their energy needs as well. Birds and mammals, however, can be looked upon as gas guzzlers of sorts because their constant high body temperatures and metabolic rates require frequent ingestion of food and a body capable of supplying high amounts of oxygen. Thus, rather than being considered lazy, reptiles should more accurately be referred to as energy conservers or fuel-efficient vertebrates.

Crocodilians and other reptiles can maintain intense anaerobic activity for only short periods of time. They are unable to carry out prolonged physical exertion because they are unable to supply enough oxygen for the aerobic metabolic process to take place. Therefore, they become quickly exhausted and need a considerable recovery time after intense physical activity. Lactic acid is an end product of anaerobic metabolism, and a buildup of large amounts in the tissues causes a drop in the blood pH. This results in lactacidosis or metabolic acidosis, a condition in which the acid/base balance of the body is altered. Oxygen debt (anoxia) is also an issue. The body needs to get back into an equilibrium or resting state, so oxygen can move back into the tissues and the pH balanced into the slightly alkaline range. Overexertion during capture can lead to death, especially by drowning, as a result of acidosis. Crocodilians normally take several hours to recover from exercise, but the recovery time may take more than a day.

The pH scale ranges from 1 to 14: below 7 is acidic, above 7 is alkaline, and 7 is neutral. The pH in vertebrates tends to be between 7.0 and 7.5. The recorded values for crocodilians are amazing for any vertebrate, as they can range from being extremely alkaline after feeding to extremely acidic after exercise. Values have been recorded at an acidic level of as low as 6.6, which would prove deadly for any other animal known. At the other extreme, after feeding, values have been noted that are close to 8.0, and there is even a recorded value above 8.0. Nearly all of the biochemical reactions in the body are influenced by the pH of the body fluids, so it is imperative that the pH stay within critical limits. Because crocodilians tend to push the boundaries of their acid-base balance, stress induced by humans can push them over these limits and make it difficult or impossible for them to get back into equilibrium.

Diving and Submergence

Diving and submergence relate closely to anaerobic respiration and metabolic rates. Diving results in reduced blood flow to the muscles while maintaining the blood flow to the heart and brain. As a result, energy expenditure is not great during diving and submergence. A common question is just how long crocodilians can stay submerged. Roland Coulson, a pioneer in studies that began in 1948 relating to the biochemistry and physiology of alligators, provided the most reliable estimate that I am aware of. He and his co-researchers believed that it was possible for an adult alligator to stay submerged for up to 12 hours at 82.5°F (28°C) and for a period of days at temperatures between 44.5°F and 50°F (7°C and 10°C). They noted that one of their experimental alligators stayed on the bottom for 6 hours at a water temperature of 80.5°F (27°C), before coming back to the surface for a period of a couple of seconds, only to dive again (see Coulson and Hernandez 1964; Coulson, Herbert, and Coulson 1989).

Adverse Conditions

Crocodilians have a variety of mechanisms to deal with adverse conditions, such as extreme temperatures and weather.

Near-Freezing/Freezing Temperatures

Freezing temperatures are typically encountered in nature only by the two species of alligators, which are the only living temperate-dwelling crocodilians. Rather than remain submerged in their dens during freezing temperatures, American alligators are known to exhibit an "icing response." This involves the alligator's nostrils protruding at the water's surface when ice is present in the water, and the rest of the body is angled down into the water. If temperatures were extremely cold and ice was not present, the alligators were found to be in their dens in the North Carolina study that first documented the "icing response" (Hagan, Smithson, and Doerr 1983; fig. 3.5).

Larry McNease, former biologist at Rockefeller Refuge in Louisiana, told me that alligators there had to be "chipped out" of the ice because ice had frozen around their snouts. Survival seemed to depend on how long they had been subjected to these harsh conditions. Why does the alligator leave the den when the water ices up outside the den? Does this

Figure 3.5. Dens can provide shelter year-round. When water levels are up, seeing an alligator's head sticking out the side of a bank may be the only indication that a den is present.

mean that the water in the den ices up as well? An alligator den tends to have a slight incline downward from the entrance to the tunnel. As a result, the back of the den should have an air space between the soil roof and the surface of the water below. If oxygen enters the den from the water/air interface outside the den, then ice outside the den would keep oxygen from entering the den, so the alligator has to go outside to breathe. It would not matter if the very back of the den did or did not have ice on the water. Another possibility is that oxygen can enter the den from the soil above, especially in cases where the den is built below a tree's root system, which may loosen the soil. Since land freezes faster than water, this would also cut an oxygen route to the den.

It might be expected that alligators become very sluggish during extremely cold weather. Such is not the case with alligators and some other reptiles. Years ago, I worked at a zoo in central Illinois during the summer and had my Nile monitor lizard on display loan there in an outdoor enclosure. Temperatures dropped unseasonably low from a high of over 82.5°F (28°C) to near freezing overnight without warning. I went to capture the lizard early the next morning to move him indoors. This animal was not tame, and the only time that he was calm enough to handle without him trying to bite and lash his tail was when he was bathing in warm water. He would remain in the water as long as it stayed warm, but would leave it as soon as it began to cool down. He would close his eyes and relax his body in the warm water, and I could actually pick him up and hold on to him for half a minute until he

became alert and began flailing. I had never dealt with this lizard or any other relatively large reptiles at temperatures near freezing, and I had expected him to be sluggish. Nothing was further from the truth! He was very agitated and lunged at me repeatedly. His behavior was even more aggressive than usual. Once I finally captured him and brought him indoors to a warm enclosure with a heat lamp, his behavior returned to normal. Large lizards have the ability to store heat, just as crocodilians do, and this allows them to be able to react even at low temperatures.

I was reminded of this incident with my monitor lizard several years later when I had several juvenile alligators in a shallow in-ground wading pool that had a clay substrate, and a similar temperature drop occurred in the fall in the Houston area. Temperatures were expected to reach the freezing mark and had already dropped to 40°F (4.5°C) in a period of a few hours, so I planned to capture the alligators and take them inside. I tried to catch the alligators in a net, as they were down in the mud so that they could not easily be seen for me to secure them by hand. Every time the net came into contact with one of the alligators, it would hiss and lunge, but I was unable capture any of them. They would stir up the mud and move to an area of the pool away from the net. I finally resorted to dripping water from a hose into the pool overnight to avoid any possibility of ice forming on the water's surface. Although these alligators were kept in a naturalistic enclosure, I used them from time to time for speaking engagements and never before had a problem with capturing them or their exhibiting any type of aggressive behavior.

An adult female alligator that displayed the icing response at a temperature between 39°F and 41°F (4°C and 5°C) in the previously mentioned North Carolina study hissed, moved toward the researchers, and vocalized with a bellow-growl. It appears then that this aggressive, agitated behavior is not unusual and may even be expected with the approach or existence of freezing temperatures. Remaining responsive and active may play a role in survival by increasing the metabolic rate and thus internal heat production (especially in adults). The stress placed on the body must be great in order to facilitate major metabolic changes in response to such a rapid change in temperature. The agitated behavior is perhaps a direct result of the stress placed on the body, a "fight or flight" response. Cooling an alligator down or the use of ice as a management tool is inappropriate and inhumane.

It is uncertain what percentage of alligator deaths resulting from near-freezing to freezing winter temperatures results directly from exposure to extreme cold or from suffocation under the ice. The fact that alligators are known to sometimes die in the icing position with their snout above water with ice around them points to cold weather alone as a cause in some cases of winterkill. However, it is unknown if there is a critical temperature (body or environmental) below which death results, a critical time duration of extremely cold temperatures beyond which survival is not possible, or a deadly combination of the two factors. It is predictable in Southeast Texas that after several days of freez-

ing temperatures, a week or two later there will be some dead alligators floating in the water. Furthermore, we know that these animals tend to range in length from 4 to 6 feet (1.2 to 1.8 m). One explanation for this size-related mortality is that larger alligators have an advantage over smaller alligators regarding thermoregulation.

However, over a period of days of extreme cold, a size factor would not likely be the lone reason for increased survival. It also does not explain why we do not find a winterkill of young alligators. Even if their carcasses would not be as discernible as those of larger alligators, we see a substantial percentage alive in the spring following a winter with observed mortality of subadults and smaller adults, proving that they can survive. Large adult alligators with well-established territories may have an advantage of a warmer microclimate. I have observed that access to areas protected from wind, especially with deeper waters, tends to be associated with the dens and "gator holes" of larger alligators at Brazos Bend State Park. Another factor that can result in warmer temperatures in deep-water areas is the presence of a substantial biomass of submerged plants. During cold winter days, there is often a lone alligator or two of large size with their heads showing or juvenile alligators with most of their dorsal body surface visible within a large mass of coontail (*Ceratophyllum demersum*). I took temperatures within several large dense stands of submerged plants and found them to be several degrees higher than that of the surrounding water. The concept of favorable microhabitats enhancing survival when it is cold may also explain the survival of alligators found outside their normal range.

Hot Temperatures and Drought

High temperatures do not seem to result in extreme stress to alligators in the wild as long as they have a deepwater area, preferably with a muddy substrate, that remains cooler than the air temperature. Heat accompanied by drought is more problematic. Brazos Bend State Park has been an ideal site to observe alligators during drought, as there are a number of water bodies with varying depths (Hayes-Odum and Jones 1993). Some areas of marginal habitat quality, such as the prairie pothole area, are exposed to drought conditions on a more frequent basis than elsewhere in the park. Some water bodies dry up partially, and others become totally dry.

There is a movement of a number of alligators of all sizes to areas that still contain deep water during drought. However, not all alligators move to water. Alligator dens that become partially or completely exposed due to declining water levels often contain an adult or pods of juveniles. When the water levels in the vicinity of a den become low enough or dry up, the perimeter of the alligator hole adjacent to the front of the den becomes apparent as a water-filled or waterless depression (fig. 3.6). At times the dens themselves may have no visible water in the entrance. When water is present in the den or alligator hole, sometimes there are fish and frogs in residence.

Adult den occupants are believed to be pri-

Figure 3.6. When water levels are low, dens and their associated ponds ("gator holes") become exposed. Dens may be built under the root system of a tree as this one is. The root systems of trees help prevent the dens from collapsing, although they sometimes do.

marily females, as they are commonly found in association with nests or have offspring with them in the den/alligator hole. I believe that at least a portion of the alligators remaining in dens during the day may make nocturnal trips to water bodies that still contain substantial water. The evidence is in the form of numerous tail drags and tracks in the den vicinities, either in the mud or on footpaths between dens and permanent water. These excursions could be for the purpose of hunting for food and/or ridding themselves of excess heat in the deeper water. All of the alligators observed during several different droughts appeared to be able to "wait out" the drought and remain in reasonably good condition. The only mortality that has been observed during dry periods is cannibalism, which especially affects the subadult and smaller adult groups. There is also indirect evidence that juvenile mortality may increase as well.

Reproduction can be affected by drought by reduction of bellowing and courtship activity, a change in the location of nesting sites, a decrease in nest number, and a decrease in mating and nesting in the next year. There can also be a decrease in parental care when an animal is exposed to prolonged periods with no water at the den/nesting site. A nest that was hatching at the prairie pothole area in the park was not opened by the female. A decision was made to remove the young before they became prey or died from heat and a lack of water. In one drought observed at the park that was coupled with a human-made drought due to draining a major water body, an increased winterkill was believed to have been induced by nonres-

ident alligators overwintering in a water body that was not their usual territory. In an alligator population at or near carrying capacity, an increased number of animals inhabiting a lake would tend to have a difficult time securing quality microhabitat that would protect them from freezing temperatures.

We observed that behavior of alligators in areas affected by drought differed according to whether or not they had access to permanent water. Adults and juveniles associated with open-water areas typically displayed normal behavior both in the water and on land. Adults in areas with little to no water became very agitated and hissed loudly when approached, while juveniles were extremely wary and submerged if there was sufficient water or moved quickly into their dens. Undisturbed alligators were inactive and nearly motionless.

Flooding

Knowledge about the behavior of alligators during flooded conditions is a bit sketchy. Whereas drought exposes alligators and their activities and allows humans greater access to them, flooding makes alligators less visible and accessible. Occasionally a displaced alligator appears in an area that it was not known to inhabit previously. Sometimes as floodwaters begin to recede, several alligators become visible at their den sites. It seems that these alligators stayed at the site or remained nearby.

A few comments can be made about how they might be affected by extremely high water levels. Since an alligator breathes using lungs, rushing waters will make it difficult for an alligator to keep its head above water to avoid drowning. The nostrils, mouth opening, and ears close automatically to seal out water only when it dives, so water would rush into these openings. Alligators would have to expend a lot of energy to fight floodwaters sweeping them away. They would also find it difficult to thermoregulate behaviorally if they could not move back and forth between land and water.

Reproduction can be affected, depending on the timing and duration of the flooding, and the evidence is easier to observe. The flooding can interrupt social behaviors and breeding, result in embryos drowning within nests, adversely affect juvenile survival, and reduce nesting the next year. In a study undertaken at the Rockefeller refuge in Louisiana, eggs taken from nests were subjected to simulated flooding during the first, third, sixth, and eighth weeks of incubation (Joanen, McNease, and Perry 1977). Some eggs of each age were immersed in water for 2, 6, 12, and 48 hours and then placed in incubators to see if they would hatch. Two hours of submersion did not seem to adversely affect the survivorship of embryos. Six hours and 12 hours of submersion resulted in about the same survivorship but less surviving young than the comparable control group (i.e., those embryos of the same age that were not subjected to flooding)— an average of 67% live hatchlings. All of the older embryos in the 12-hour flooding died. The 48-hour submersion treatment had only three survivors at one week of incubation. At Brazos Bend State Park, the relatively flat prairie pothole area is routinely subjected to flooding conditions as well as drought, and one nest partially flooded before the eggs were removed by park personnel and incubated artificially.

Hurricanes

So what do alligators do in a hurricane? Hurricane Ike made landfall near Galveston on September 13, 2009, and gave us a glimpse of what effect hurricanes can have on alligators. Drowning occurs during the storm surge when these animals cannot keep their heads above water. Flying debris can result in their injury or death. Alligators can even be swept away during the storm surge and end up in a locale some distance from their normal home range. An extreme example of this was documented in which a live juvenile alligator was found amid debris on the beach at Padre Island National Seashore on September 28 after the hurricane (Elsey and Aldrich 2009). It was dehydrated and emaciated. From the presence of an ID tag on the webbing of each hind foot and tail notches, it was determined that it had been marked as part of study by Louisiana Department of Wildlife and Fisheries. Six weeks prior to the storm, it had been released in Johnson's Bayou, Louisiana, which is located nearly 304 miles (489 km) from the beach where the alligator washed up. But the post-storm conditions can have a huge impact as well. Lack of fresh water caused dehydration and eventual death to many alligators in coastal areas. See chapter 4 for an account of alligator rescues after Ike.

Alligator Food Habits in the Wild

What do alligators eat in the wild? Alligators are carnivorous, or meat eating, at all life stages. They eat a wide variety of animals, the types and species of which are dependent mainly on the availability, the size of the alligator, and size of the prey item. It is ironic that predators of hatchling alligators, such as bullfrogs, herons, and egrets, in turn become prey of adult alligators. It has often been said that alligators eat anything that moves. I can attest to the fact that they also eat prey that has long since stopped moving. A free-ranging adult alligator was offered a dead ratsnake and readily ate it. I have also known adult alligators to eat flattened roadkill directly off the pavement.

Nonfood Stomach Contents

Alligators often eat items considered to be nonfood. Commonly occurring plant material in their diet results from swallowing it intentionally or accidentally, especially while ingesting prey; secondarily ingesting it via stomach contents from an omnivorous (both meat- and plant-eating) or herbivorous (plant-eating) prey item; and occasionally picking up a stick, a piece of plant tuber, or other relatively large object and eating it. However, they may derive nutrition from the ingested plant material, and it may be a necessary part of their diet to stay healthy.

A few juvenile alligators that I have examined have had small rocks along with the food items that I pumped from their stomachs. One juvenile possessed 20 rocks in its stomach. Another juvenile's stomach contained a small piece of lumber. I had access to the entire stomachs taken from harvested adult alligators and was able to look at food and nonfood items found in them. They included rocks, pieces of lumber, broken glass, a balloon, shotgun shells, a bottle cap, a piece of metal, hairballs, a wad of monofilament fishing line, and a concretion.

The hairballs may have been in the stomachs of nutria eaten by the alligators or may have formed within the stomach of the alligator. The source of the soft rocklike concretion is unknown. It was very distinct from the pieces of wood and plant tubers inside these alligators. E. A. McIlhenny (1935) found similar objects in the stomachs of alligators that were taken from their dens in February and March but never from alligators killed later in the year when they were actively feeding and digesting their food. He theorized that these objects were food remnants that were not digested after the alligators began their period of winter inactivity. He thought that the muscular action of the stomach molded them into a solid brown woodlike mass. However, the concretion that I noted was found in a stomach during the hunting season in September when alligators are known to be feeding.

Some have speculated that these are gastroliths (stomach stones) to be used as "grinding" stones, as birds use to grind food in their gizzards, or as ballast to aid in controlling buoyancy when in water and helping the animal remain submerged. Whichever explanation proves to be the reason behind gastroliths, both wild and captive animals purposely swallow such objects. A couple of adult alligator stomachs I examined contained mostly nonfood items, and I referred to them as "garbage can" alligators. These animals appeared to be in good health, and it is unknown if they ate fewer food items overall than other alligators.

Diet of Texas Juvenile Alligators

The most common dietary items of juvenile alligators (< 4 feet, or 1.2 m) at Murphree

WMA and Brazos Bend that I discovered when pumping their stomachs (see "Capturing Alligators to See What They Eat") were insects and insect larvae in varying stages of disembodiment and small crustaceans such as crayfish, freshwater shrimp, and seed shrimp. Snails, spiders, small fish, and a rare snake, small bird, or mammal were found less frequently in the stomach contents. Only one of the alligators had an empty stomach.

Insects were found in 98.9% of the 187 juvenile alligators whose stomachs I pumped (the frequency of occurrence) and accounted for 92.8% of the total number of food items present (the percent composition). A single animal was found to contain as many as 59 insects, but the mean number of insects per alligator was 9.6. The adult insects most often found were giant water bugs, water scorpions, creeping water bugs, water scavenger beetles, predaceous diving beetles, and scarab beetles. Most common larval insects were horseflies, rat-tailed maggots, dragonflies, and predaceous diving beetles. The common insects were observed in swarms (either swimming or flying) at both study sites and might have provided opportunistic situations in which young alligators could eat large numbers of a single species within minutes. This scenario was supported by the presence of a high number of a particular insect species in individual alligators with all the insects in the same state of decomposition.

Crustaceans were eaten by 23% of the juveniles but accounted for only 3.6% of the total food items encountered. Mollusks (snails) occurred in 11.8% of the alligators' stomachs

and made up 1.7% of the total food items. Arachnids (spiders) were the least common invertebrate noted, as they were found in only 3.2% of the alligators' stomachs and made up less than 1% of the total food items. This finding is probably a reflection of the fact that most spiders are terrestrial and are not as common as the huge number of insect species.

Fish were the most common vertebrate found in juvenile stomach contents. Minnows, killifish, sunfish, and mosquitofish were most often encountered in this study, with mosquitofish being the most common. A total of 10 alligators, or 5.3%, contained fish, and fish contributed just under 1% of the total dietary items.

The remaining vertebrate groups combined made up less than 1% of the total food items. No amphibians were found in stomach contents. A single snake was the only reptile found, as evidenced by belly plates discovered in one of the largest juveniles sampled, with a length of 3.96 feet (1.20 m). Five alligators contained bird feathers. The numerous feathers in each case suggested that an entire bird was ingested rather than just feathers. Seven alligators contained hair and skeletal fragments of what appeared to be a maximum of one mammal each. A rabbit and cotton rat were identified by hair and bone samples, but the five other mammals remained unidentified. Due to the lack of a carcass or intact skeletal components, the size of the prey when intact and how much was consumed were not determined.

I pumped the stomachs of 155 juvenile alligators from the Murphree WMA (coastal habitat) and 32 juvenile alligators from Brazos Bend State Park (inland habitat). There were significant differences between study sites. Brazos Bend had significantly higher values for both the frequency of occurrence and the percent composition for fish, crustaceans, and mollusks, but Murphree WMA had higher numbers for adult insects. However, Brazos Bend had a greater diversity of insect species than Murphree WMA. Given that Brazos Bend is a prime inland habitat with a much greater diversity of plant species and a greater interface of land and water than coastal habitat, it is not surprising that it was more diverse in terms of the species present for alligators to eat.

Diet of Texas Adult Alligators

I used the stomachs of adult alligators taken by hunters in coastal areas from Jefferson, Chambers, and Orange Counties to ascertain the dietary habits of adults in coastal habitat. The alligators ranged in size from 5 feet to 10 feet, 9 inches (1.52 to 3.28 m) in total length with a mean length of 7 feet, 9 inches (2.36 m). Of the 42 stomachs examined, two were empty and five contained only nonfood items. The most notable difference in the food habits of adult versus juvenile alligators in Texas was the dietary shift from invertebrates to vertebrates. The frequency of occurrence was 104.7% versus 45.2%, and the percent composition was 71.4% versus 28.5%, with vertebrates having the higher values. Mammals were the most common food item in terms of both frequency of occurrence and percent composition (38.1% and 24.3%). The invertebrate group of crustaceans came in second with values for the frequency of occurrence and percent composition

of 33.3% and 21.4%. The descending order for the frequency of occurrence and percent composition of the remaining groups was reptiles, amphibians, fish, birds, mollusks, and insects.

Muskrat (*Ondatra zibethicus*) and the nonnative nutria (*Myocastor coypus*) were the mammal species identified by bones and/or hair. Unidentifiable mammalian skeletal remains were found as well. Several purple gallinule (*Porphyrio martinicus*) were present in alligator stomachs, plus unidentified feathers found in one stomach. A glossy water snake (*Regina rigida*) and green water snakes (*Nerodia cyclopion*) were found nearly intact. Snake ventral scutes and a snake skeleton of undetermined species were encountered also. The red-eared slider (*Trachemys scripta elegans*) was identified in stomachs in the form of a skeleton or scutes, and additional turtle scutes could not be positively identified but were possibly the same species. Large frogs belonging to the family Ranidae were believed to be the locally abundant pig frog (*Rana grylio*) rather than the bullfrog (*R. [Lithobates] catesbeiana*). Fish species detected were crappie (*Pomoxis* spp.), buffalo (*Ictiobus* spp.), and shad (*Dorosoma* spp.), as well as several unidentifiable fish.

In terms of invertebrate species identified, crustaceans represented were crayfish (*Procambarus* spp.) and blue crab (*Callinectes sapidus*). Mollusks were present in the form of single, small snails, which may have been accidentally ingested during feeding. One insect was found, a large tabanid (horsefly) larva in the stomach of a small female alligator (6 feet, or 1.8 m in length). Given the excellent preservation state of the larva, its large size, and the relatively

Capturing Alligators to Learn What They Eat

It was hot (77°F to more than 86°F, or 25°C to more than 30°C) when we started out in the airboat near dusk to begin a night of captures at the J. D. Murphree WMA in Port Arthur. I was always amazed that in a span of a minute I could be freezing when we got to open water, where we hit high speeds to quickly get to the first capture site. When we reached a capture area, we slowed down from a "roar" to a "putt-putt" sound, and the air around me once again became hot and muggy. The TPWD staff member piloting the airboat, usually Perry Smith or Lee Ann Johnson, shined a spotlight through the water until we located a pod of juveniles. Sometimes I had one to two assistants helping with the captures. We got on our knees on top of the bow of the boat and reached over the front and sides to grab juveniles behind the head. If there were numerous animals in one spot, we ended up with an alligator in each hand. On extremely successful nights, we captured close to 100 young alligators. In order to keep alligators from different areas from getting mixed up, we tied surveyor's flagging in the color that represented a particular capture area around each alligator's body just in front of

its hind legs. The snouts were taped shut with masking tape to ensure that they did not injure one another during transport. The alligators were placed into ice chests and wooden duck boxes to transport them to the work area where I would collect individual data.

When we reached the work area onshore, each group of animals was placed in a metal holding tank containing shallow water. Each individual was fitted with consecutively numbered size 1 Monel self-piercing aluminum tags on the web between the second and third digit of each hind foot. I was advised by Florida alligator researcher Jim Kushlan to "double-tag" the alligators in case of tag loss. By the end of the study, I had not experienced any tag loss in those alligators recaptured one or more times. However, double tagging proved invaluable, as a substantial number of tags became discolored and even corroded on recaptured animals, and the numbers were sometimes partially illegible on one or both tags. We recorded total length of the alligator and the area it came from. If we had a large number of animals to process, it sometimes took until 4:00 a.m. to finish tagging and measuring. It was imperative to get the animals tagged as soon as possible for humanitarian reasons, so that if we accidentally had an escapee, it would be missed and subsequently found. We then pumped the alligators' stomachs to determine their diet using the method of Taylor, Webb, and Magnusson (1978).

small size of the alligator, it may have been taken intentionally rather than being ingested secondarily or eaten incidentally in the process of grabbing prey.

Since I have spent a considerable amount

of time in the field at Brazos Bend State Park, I have had the opportunity to observe alligators take a variety of prey species. Several alligator carcasses found through the years at Brazos Bend also have provided an indication of food habits through stomach contents. As there is no hunting at Brazos Bend, I have compiled a list of animals known to have been eaten by adult alligators there. The notable distinction between the stomach contents from the coastal areas and the known prey items from this inland site is the inclusion of terrestrial animals in the alligator diet at Brazos Bend. Nutria, white-tailed deer (*Odocoileus virginianus*), raccoon (*Procyon lotor*), armadillo (*Dasypus novemcinctus*), swamp rabbit (*Sylvilagus aquaticus*), and feral pig (*Sus scrofa*) are mammals known to have been preyed upon. Two dogs (*Canis familiaris*), a poodle and a Weimeraner, belonging to park visitors were taken by alligators shortly after the park opened to the public after their owners tossed a stick and a Frisbee, respectively, into the water for the dogs to fetch. Philippe observed a near miss of a coyote (*Canis latrans*), the last of a pack of more than 20 individuals crossing Big Creek, being taken by an alligator.

Numerous bird species have been observed being eaten by alligators at Brazos Bend, including great egret (*Ardea alba*), cattle egret (*Bubulcus ibis*), common moorhen (*Gallinula chloropus*), American coot (*Fulica americana*), and black vulture (*Coragyps atratus*). The black vulture was taken as carrion. Red-eared slider, spiny softshell (*Apalone spiniferus*), and Texas ratsnake (*Elaphe obsolete*) were the reptile species known to have been consumed. Subadult and small adult alligators were sometimes killed or killed and eaten by larger alligators. Fish caught by alligators and eaten included catfish (*Ictalurus* spp.), largemouth bass (*Micropterus salmoides*), and alligator gar (*Atractosteus spatula*). In addition to food items, nonfood items were found in the stomachs examined from dead alligators at Brazos Bend, including sticks that were smooth and worn, several fishing bobbers, a shotgun shell, and a beer can.

Although observations of prey being eaten at Brazos Bend do not correspond to what we found in the stomach contents from Murphree WMA, it is evident that adults at an inland site relied on vertebrate prey as did the adults at the coastal site. Terrestrial vertebrates are commonly eaten at the inland site, just as the juvenile diet at the inland site represented terrestrial species in the form of invertebrates.

Comparison of Alligator Diet Studies in Texas and Other States

The diet of alligators in Texas seems to be similar to that of Louisiana alligators. Florida has some prey species that are not present in Texas or Louisiana, such as the apple snail (*Pomacea paludosa*), which is important to the diet of Florida alligators, and horseshoe crab (*Limulus polyphemus*). (There are some introduced apple snails in Texas, but they are in localized areas and not likely to be a food item of very many alligators.) This is the first study in which frogs were a relatively important food item of adults. Juveniles of all sizes in my study were dependent on insects as the mainstay of their diet. Few other studies included small juveniles. Platt, Brantley, and Hastings (1990) pumped stomachs of 101 juvenile alligators of all sizes in southeastern Louisiana and found that crustaceans, insects, and small fish made up the bulk of their diet. The species were those commonly encountered in the Texas study I conducted. Crustaceans were the most important dietary item. Delany (1990) looked at 80

A Field Note from Philippe on Prey Capture

Alligators are able to use their body as a tool to catch fish. Although alligators seize fish when swimming in the deep waters of a bayou or a lake, they also capture fish in shallow waters, using their body like a seine net to trap the fish. I filmed this rarely seen behavior in Brazos Bend State Park during a drought when the water level in a bayou close to Pilant Slough was less than a foot deep. The alligator arrived at the bayou early in the morning and stretched his body across the width of the bayou so that he could touch both banks. Fish swimming downstream were stopped by the alligator's body. Some fish jumped over the alligator to escape and were able to continue their course downstream. Others did not try and became easy prey. When the alligator had enough fish trapped close to his body, he bent his tail toward his mouth, making a sort of fence. Then he could catch many fish by moving his mouth quickly in this fish pond.

I also watched a large alligator that was blind in one eye prey on turtles that came onto the trails at Brazos Bend State Park to lay their eggs. I saw him catch six adult turtles in a two-week period on a stretch of trail that is a favorite spot for turtles to lay their eggs each year. This alligator probably knew this location was a popular nesting area and came here every morning. He lied down close to the trail to watch for turtles. As soon as he spotted a turtle, he went in the water and swam close to the spot where he anticipated that the turtle would return to the water. As the turtle entered the water, the alligator grabbed the turtle and swallowed it whole after he crushed the shell. The alligator, who died in summer 2004 of unknown causes, appeared to be able to see turtles over 50 yards (45 m) away with his good eye. He used ingenuity to catch prey, overcoming the disability of his blind eye that would hinder him in open water.

juvenile alligator stomachs from a freshwater lake in Florida and found insects to be the most important part of the diet of juveniles less than 2 feet long (0.6 m), but with increasing alligator size, insects declined in occurrence more than other food groups, including fish, other vertebrates, snails, and crayfish.

Few of the alligators sampled in this study of Texas alligators were in the size class of subadults (4 to 6 feet, or 1.2 to 1.8 m in length). This is the size class that undergoes a dietary shift in which vertebrates start to replace invertebrates in importance (Brantley 1989; Delany and Abercrombie 1986; McNease and Joanen 1977; Wolfe, Bradshaw, and Chabreck 1987). It has been found that a change in skull and dental size, shape, and proportions with growth may be an adaptation for capturing and feeding upon large prey (see Erickson, Lappin, and Vliet 2003 for a review of the literature on this topic). For example, adult turtles become a dietary item of alligators once they reach 8 feet (2.4 m) in length. The percent composition that a larger food item such as crayfish makes up is misleading unless it is also considered in terms of how much more energy it provides than smaller prey.

Shortcomings of Diet Studies

There are numerous pitfalls when considering stomach contents to be a totally accurate depiction of what a crocodilian has eaten at any given time. Most important, some items digest rapidly, and others tend to persist for long periods of time. The crocodilian stomach is the most acidic recorded for any vertebrate, which enables bone to be digested. Frogs are known to be digested within a period of three days. I have watched a complete young sunfish disintegrate in my hands after being pumped out of a juvenile alligator's stomach. I have also seen many chitonous insect parts that were in small pieces inside the stomachs of juveniles. These tend to remain in the stomach for relatively long periods of time. It has even been suggested that these chitonous remains act as gastroliths for grinding food. Artifacts such as hair, feathers, and snake belly scales that I have found in stomachs show that these vertebrates have been eaten, but I am unsure how long they can remain after the rest of the animal has been digested.

An alligator could eat several crayfish, yet one adult nutria would contribute more nutrition. To make the importance of any given food item relative in comparison to another, Chabreck (1971) and other researchers have suggested using live weights to account for the differential sizes of various prey species. I do not agree that this is a valid method, as this assumes that an entire carcass of a particular species is ingested. Since alligators eat carrion (dead, decaying animals) and more than one alligator may eat a large prey item, this can skew the findings unless enough of the animal is left to ascertain that the entire animal was eaten.

Then how does one interpret partial remains? They can be discounted, in which case numbers of animals eaten and entire species may not be utilized in the final tally. It appears then that there is no "perfect" way to make the stomach contents equitable. Perhaps the best way to deal with the data is to list exactly what species are in the stomach, how much is left, what form it is in, the frequency of occurrence (what percentage of stomachs contained a particular taxonomic category), and the percent composition (what percentage of the total food items was that particular taxon). Additional knowledge, such as accounts of predation or scavenging with the species being taken, computation of seasonal changes in stomach contents, and factoring in differential rates of digestion for each prey species can keep the relative importance of a particular species or other taxon in perspective.

Methods of Predation and Food Procurement

Alligators range from active predators, to sit-and-wait predators, opportunistic predators, group hunters, and scavengers. They are best known for lunging at and grabbing prey, either in the water or on land near the water (fig. 3.7). As sit-and-wait predators, they can quietly lie submerged at the water's edge and grab

Figures 3.7. (a)
The alligator
is known as a
predator that
lunges at and
(b) grabs prey in
the water or at
the water-land
interface.

a mammal as it comes to the water to drink. They can take advantage of a situation that presents itself, such as water flowing through the pipe of a water control structure from one water body to another, and seize crayfish or slow-moving fish as they come out of the pipe.

Philippe has filmed a feeding behavior of an alligator catching fish by using its body as a seining net of sorts to block fish and then move its tail (and the fish along with it) toward its mouth so that it can eat the fish. This behavior has been mentioned in the literature before (Graham and Beard 1973), but Philippe may be the first to film it (fig. 3.8).

Dennis Jones of TPWD and I (1990) described a "tail-wagging" behavior in alliga-tors that appeared to be associated with feeding (fig. 3.9). The posterior third of the tail rapidly moved through the water and mud in a 180° arc. Simultaneously, the head usually moved from side to side in a 90° arc, often with the mouth held partially open. Occasionally an alli-gator would lunge and then appear to swallow something. We postulated that this behavior forced prey such as fish and amphibians up to the surface for air after the surrounding water was stirred up and brought the prey closer to the head for ingestion. This may be a variation of the body net-fishing technique.

Since publishing our observations over two decades ago, we are aware of two lines of research that augment the knowledge of how

Figure 3.8. An alligator can use its body as a net to trap fish and then sweep its tail and the fish toward its mouth.

Figure 3.9. A tail-wagging behavior that Dennis Jones and I have observed that appears to be associated with feeding. The alligator wags its tail back and forth, which may function to stir the water up and force fish and other prey items to the surface for air and bring them closer to the mouth. The alligator moves its head from side to side with mouth held partially open and appears to swallow something from time to time.

alligators locate food. Paul Weldon, formerly a biology professor at Texas A&M University, and his students found that alligators can detect both water- and airborne chemicals from meat and that they may use chemoreception to locate food both on land and underwater (Scott and Weldon 1990; Weldon et al. 1990). Daphne Soares (2002) discovered that dome pressure receptors, a series of pigmented, pinhead-sized dots on the faces and inside the mouths of crocodilians, indicate that crocodilians can sense ripples in the water. The receptors work only when the animals' heads are partially submerged in water, not when they are on land or totally submerged. They are believed to function in prey detection.

Philippe noted on several occasions an alligator, which was blind in one eye, waiting for turtles to come out on land to lay their eggs.

The alligator then waited at the spot in the water where it expected the turtle to return. The turtles were easy prey for the alligator as they entered the water (fig. 3.10).

Although I agree with Philippe that other alligators probably take advantage of nesting turtles, I think that this alligator had to exploit such situations due to its partial blindness, as it would be at a disadvantage seizing prey if the animal escaped to its blind side. This demonstrates that a crocodilian that is feeble or disabled can have some control over obtaining prey to greatly increase the likelihood of success. For crocodilians overall, it provides an example of habitual behavior of prey species being used to the advantage of the crocodilian

predator. It has been asserted that some individual crocodiles of large, aggressive species have resorted to hunting humans as an easy prey source when they were no longer capable of obtaining more challenging prey. While such accounts are often misreported or exaggerated, there is probably some truth in them.

The head and tail have been known to be used as tools or weapons of sorts to dislodge, flip, or give hammerlike blows to prey. For example, crocodilians can use their heads to deliver side-to-side blows to stun, break bones, dismember, or knock prey into the water (fig. 3.11).

Their tails have been known to strike or dislodge prey. In bird rookeries that have nests

Figure 3.10. Philippe observed on several occasions an alligator that was blind in one eye preying on turtles that had come out of the water to lay their eggs.

a

b

Figure 3.11. (a, b) The alligator's head, teeth, and jaws are powerful weapons against prey.

then swallow the birds. A bird-watcher, Margaret Jones (1988), chronicled this behavior at Brazos Bend State Park. There is a photo of an alligator trying to grab a young bird in this manner in Pooley (1989, 87).

Observations of crocodilians with floating sticks above their snouts, swimming through the water with the rest of their body submerged have been made, and those sticks are believed to be tools used by the crocodilians to camouflage themselves during hunting (Dinets and Brueggen 2015). I noted one instance of this behavior at Brazos Bend at Pilant Slough, which has a lot of floating debris in the water. The alligator caught my attention by having sticks and twigs on top of its head and making no apparent attempt to shed them. However, I attached no particular significance to the behavior until I read about it relating to hunting for prey.

Crocodilians can hunt for prey collectively, that is, two or more animals can block or direct prey such as fish so that the prey does not escape and all participants can feed. For example, crocodilians can form a semicircle or line to keep fish from passing through the line to another area. Each animal keeps its place in line and nabs fish as they go through. In 1999 at Brazos Bend State Park I filmed a group of several alligators lined up on the south side of Elm Lake between the south-central islands and the shoreline. It was August and hot. The fish were oxygen deprived and congregating in this open-water area to gulp air at the surface. The alligators formed the line, and occasionally I could see one grab a fish. Texas alligator farmer Mark Porter recalls fishing one night

of aquatic birds, such as herons and egrets, alligators not only wait for young birds to drop out of nests but also raise their bodies out of the water all the way to their front legs and push against the trees to dislodge the young birds into the water. They also lunge and grab portions of trees containing nests and baby birds, bringing everything down they can grab, and with friends and realizing that some alligators were involved in a group fishing expedition as well. Porter's group stopped fishing and just sat back and watched the sight in front of them. The alligators were blocking and directing fish to an area where they could be more easily caught.

Crocodilians are also known to engage in group feeding of large prey animals such as a hippo. Large, tough prey is more easily torn apart by a group effort and would be hard for a single animal to defend. However, sometimes fights do ensue. Philippe filmed a fight over an alligator that had been killed and left to rot so that it could be more easily torn apart. One alligator was observed feeding on the carcass, and another came up and tried to take it away. On another occasion, a female alligator killed a subadult alligator and ate the front half. The posterior half was then placed in the water over a log so that it could be eaten later on. A male alligator that had mated with her while she was in the process of feeding on the kill, stole the remaining carcass 24 hours later. Perhaps the largest prey eaten by alligators, such as a deer, is not large and tough enough to make it desirable for an alligator to willingly share with other alligators.

Health

Do alligators get sick? In general, alligators and other crocodilians are in overall good health in the wild. Most serious health problems tend to be associated with captive animals. Poor diet, cleanliness issues, improper temperature, too small an enclosure, and exposure to exotic dis-

eases are all concerns in a captive environment (for information concerning crocodilian health, see Huchzermeyer 2003; Jacobson 1989).

Infectious Diseases

Infectious diseases found in wild and/or captive crocodilians include those of viral or bacterial origin, fungi, and parasitic protists (which are microscopic) or worms/wormlike animals (which can be seen with the naked eye).

VIRUSES

The first known virus to adversely affect crocodilians was discovered in captive caimans and subsequently in farmed Nile crocodiles in Africa. The virus was named "caiman pox" and was manifested in the form of circular, grayish-white skin lesions on the body. This seems to be a disease associated with captivity and has not yet been reported in alligators. However, an American alligator was found to show a susceptibility to EEE (eastern equine encephalitis), as it developed antibodies against it after natural exposure. Most recently, West Nile virus infected juvenile alligators on farms in several states beginning in 2001, resulting in the deaths of some alligators and at least one related human case of fever. Long-lived reptiles, such as crocodilians, are thought to serve as reservoirs for viruses just as birds and mammals do, even though they may not display symptoms.

BACTERIA

Bacterial cultures can be collected from skin surfaces of alligators, but potentially hazardous bacterial types rarely cause problems in wild alligators, even at the site of an injury. Bacterial problems tend to be associated with unclean captive conditions, improper temperatures, and poor basking sites. However, there is a report of the bacterial species *Aeromonas hydrophilia* and *A. shigelloides* believed to be involved in the death of American alligators in Florida in a lake with poor water quality.

FUNGI

It is not uncommon to find one or more fungus species in association with a wild alligator's skin but unlikely that it causes any problems. In captive animals, fungi can invade external wound sites and commonly affect the respiratory system, sometimes causing pneumonia. A variety of fungus species have been implicated with disease in crocodilians, including the well-known *Candida albicans* and *Aspergillus* sp.

PROTOZOAN PARASITES

The protozoans are usually thought of as the single-celled living organisms that we examine under a microscope. They belong to the former kingdom Protista (now subdivided into a number of kingdoms), which also includes the algae and slime molds. Most reptiles can harbor various species of pathogenic protozoans in both wild and captive situations, but symptoms may not always be present. For example, amoebae rarely cause problems in crocodilians. Young animals in crowded captive conditions tend to be the most susceptible to suffering deleterious effects from protozoans, as well as animals that do not have the appropriate temperature gradient to properly thermoregulate (e.g., coccidiosis has been known to cause death in young

crocodilian species in captivity). The presence and identification of protozoans are usually ascertained by an examination of a fecal smear under the microscope, but infections tend to be difficult to treat. Symptoms of protozoan infections vary according to the type of protozoan and the body system affected. Common symptoms include weight loss, a decrease in appetite, listlessness, and stools that are abnormal in some way, such as being too hard or too loose or including blood and/or mucus.

PARASITIC WORMS

Parasitic worms or wormlike parasites found in alligators include leeches, roundworms, flukes, tapeworms, and pentastomes. Leeches are not uncommon on external body surfaces or inside the mouth of both adult and juvenile wild alligators. I found one or more leeches (*Placobdella* sp.) on many of the juvenile alligators that I caught both at Murphree WMA and Brazos Bend. Occasionally they caused holes in the webbing of a foot or in the top of a caudal (tail) scute. One was attached to the tissue at the corner of the eye. Common attachment sites were on the upper sides of the body, on the underside of the jaw, and in the mouth.

Leeches attach to the alligator by suckers and need moisture to keep from drying out. I have noticed that once an alligator is out of water and its skin begins to dry out that any leeches present will migrate to a wet patch of skin. If the alligator becomes totally dry, the leeches desiccate and die. It has been suggested that basking behavior, including opening the mouth while basking, helps kill bacteria and ectoparasites (parasites found on the outside of the body) such as leeches. My observations tend to support this idea. Aside from basking, scratching with the claws of the feet may be a means for alligators to remove leeches. I found many leeches arranged in a whorl pattern on each side of a large juvenile's body where the hind legs attached. All four feet of this alligator were mutilated, and I think that the number of leeches increased because the alligator could not scratch them off and the area may have retained enough moisture when basking to keep the leeches sufficiently hydrated. Another potential means for alligators to rid themselves of leeches may be to rub their bodies against objects.

Tim Scott of Texas A&M University and colleagues looked at endoparasites (internal parasites) of 75 alligators from Southeast Texas (at and near Murphree WMA) and southwestern Louisiana. They created a key for the nematode, trematode, and pentastome parasites that they identified in these alligators. They found three roundworm species: *Dujardinascaris waltoni, Brevimulticaecum baylisi,* and *B. tenuicolle* (see Scott, Simcik, and Craig 1999). I found roundworms in the stomachs of both harvested adults (10 out of 54) and in juveniles whose stomachs I pumped (at least 54 out of 187). It is possible that the number of juveniles possessing roundworms was higher than that, as some could have remained in the stomachs and not been flushed out. Also notable was that there was a higher percentage of juveniles having roundworms from the inland site of Brazos Bend than the coastal site of Murphree WMA.

Trematodes are commonly known as flukes. They possess a digestive tract and are parasites of internal organs. The trematodes found in the Southeast Texas/Louisiana study included eight species: *Acanthostomum coronarium, A. loossi, A. pavidum, Archaeodiplostomum acetabulata, Crocodilicola pseudostoma, Dracovermis occidentalis, Polycotyle ornata,* and *Pseudocrocodilicola georgiana.* I did not notice any flukes in association with harvested alligators that I examined, but I could have missed seeing them since I was interested primarily in the stomachs and their contents.

However, I did find cestodes, a type of flatworm, inside one of the harvested alligators' stomachs in the form of many segments. Cestodes are commonly known as tapeworms and consist of many segments called proglottids attached to a head equipped with hooks and suckers. Tapeworms do not have digestive tracts and absorb their food directly through body surfaces. They can grow to several yards/meters in length before the mature proglottids containing eggs and sperm break off and are eliminated from the digestive tract. The eggs are then ingested by another species of animal that serves as host to the larval stage after the eggs hatch. I am unaware of any studies reporting tapeworms in crocodilians. Had I been aware of that at the time, I would have preserved the proglottids to have the species identified. It is also unclear whether the tapeworm was a parasite of the alligator or of some prey it ingested. Conceivably, a prey item containing a tapeworm could be fully digested yet the tapeworm segments remain intact due to the protective cuticle (outside covering), which is able to withstand acidic conditions within the digestive tract.

Pentastomes are wormlike internal parasites of vertebrate respiratory systems. Their name literally means "five mouths," as there are four projections surrounding the mouth that were once thought to be additional mouths. The Southeast Texas/Louisiana study found one species, *Sebekia mississippiensis*. I noticed two harvested alligators that contained a pentastome associated with the lungs.

Injuries

Injuries in wild alligators seem to heal very quickly, and there is often little to no scarring where cuts or gashes have occurred. For example, several juveniles in my mark-recapture study had laparoscopies performed on them in which a small incision was made on one side of the body in an attempt to determine their sex. A month later the site of the incision was not visible. A small juvenile ripped its venter (belly skin) open on a holding cage and also healed within a month without scarring. Even a large juvenile found in Port Arthur and released at Murphree WMA in overall poor condition, thin and dehydrated with an injured jaw, was later recaptured with the jaw nearly healed and was a healthy weight. Based on my observations, young alligators appear to have excellent healing capabilities, and wound infection does not seem to be a problem.

Adults that engage in battles seem to heal as well, except in extreme cases (fig. 3.12). This alligator was on a bank at Elm Lake adjacent to a footpath when we spotted him with injuries that were believed to be fatal. It appeared that he was attacked by another alligator on the softer skin on the side of his body. The large

area affected went through skin and muscle, leaving a hole in his body with ribs showing, and exposing the internal organs to the elements.

This photograph that Philippe caught right after a fight (fig. 3.13) demonstrates that even the tough dorsal armor of a large adult can bleed. Aside from possible injuries due to human-made structures or getting hit by cars, fights with other alligators are the primary cause of trauma for these large animals. In adults with a missing limb or a stub tail that has healed long ago, it is difficult to discern whether the injury occurred when the alligator was an adult or juvenile. It would be interesting to know in such cases what percentage of these were juveniles that were able to survive to adulthood with such limitations.

A number of scientists are conducting investigations on the immune systems of alligators, most notably Mark Merchant of McNeese State University in Louisiana and his research team. They extracted active proteins from alligator white blood cells and found that they could kill a number of types of bacteria, fungi, and viruses. Furthermore, they discovered that a successful immune response of alligators does not require them to have been previously exposed to the pathogen. Among the pathogens that alligator blood proteins have successfully destroyed are HIV and *Herpes simplex*, *Candida* yeast (a fungus), and MSRA bacteria. These amazing results hold the promise that pharmaceuticals can be developed from alligator blood proteins that could be used to treat humans for viral, fungal, and bacterial infections. Do other crocodilians have the same super immune sys-

Figure 3.12. The extreme injuries of this alligator are probably the result of a fight with another alligator. Since the body cavity was exposed, the alligator was not buoyant and could not swim. It disappeared and was presumed to have died. (Photo by Ashley Rosa)

tems? Merchant has teamed up with Adam Britton, a researcher in Australia, to test additional crocodilian species to find that out.

Anomalies

Anomalies caused by injuries, congenital differences/defects, and tumors are seen in both wild and captive animals.

DAMAGE TO CLAWS, TOES, AND TAIL TIP

Broken toes, excised claws, and loss of the tail tip were so common that they were not even notable in the juveniles in my mark-recapture study, as well as adults captured at Brazos Bend and various harvested alligators that I saw.

I believe that these injuries are often caused during "feeding frenzies" by juveniles within a pod when they go after a school of small fish or swarm of insects or insect larvae moving through the water. I discovered such injuries after feeding skirmishes in a group of four captive young alligators that I kept in a seminatural enclosure, and I observed similar altercations among wild juveniles while feeding that could result in the same types of injuries.

Figure 3.13. The tough armored plates on the back of the alligator are vascularized and do bleed. This individual's injuries resulted from a fight with another male.

MUTILATIONS TO THE FEET AND LIMBS

I noted missing toes, feet, or entire legs in several juveniles. I previously mentioned a large juvenile that had four mutilated feet, possibly from frostbite, that had whorls of leeches in an area where it could not scratch them off with its feet. This juvenile was recaptured several times due to its slowness of movement. Missing feet or limbs tend to be less common in adults than juveniles. Such physical disabilities can pose a significant disadvantage when the animal reaches the subadult size and make the alligator more at risk for cannibalism by larger alligators and impede its ability to hold a quality territory.

ANOMALIES TO THE TAIL

It is more common to see the end of a tail missing from an adult alligator than a foot or limb. Extremely short stub tails that adversely affect swimming are rare. In juveniles I have seen an extremely scarred tail in an individual less than one year of age and a short stub tail. Regeneration is possible in young animals if the remaining tail is not extremely short, and I have seen two juveniles with regenerated tails. In *The Last of the Ruling Reptiles*, Neill (1971) includes a photo of a captive adult alligator that lost most of its tail as a hatchling, and no regeneration occurred. He also includes a photo of a juvenile with a regenerated tail that had lost only a few inches of the original one. Figure 3.14 shows a large juvenile alligator from Brazos Bend that appears to have a regenerated tail. It lost over half the length of its tail and is probably at close to the limit for its ability to regenerate a new one. Regeneration may take place relatively quickly, as I found a juvenile only 19.9 inches (50.5 cm) in total length that had regenerated a tail at Murphree WMA, which means that even if this animal was injured at only a few days of age, it would have taken a maximum of a year and a half for regeneration to take place.

TOOTHLESSNESS AND JAW DAMAGE

Normally tooth replacement occurs throughout life. If a skull still containing teeth is examined, smaller teeth can be seen above the upper teeth and below the lower teeth, waiting to drop in if tooth loss occurs. Loss of teeth or jaw damage can occur as a result of feeding trauma or injuries from other alligators. Teeth can shatter or be pulled from the sockets, causing damage so that the tooth is not replaced. Most adults will have at least a couple of teeth that are permanently missing. It has been suggested that captive animals may lose teeth as a result of poor diet. Studies do not seem to bear this out. However, teeth of some captive animals do not seem to have a normal color, and sometimes the overall appearance is different from those of their wild counterparts. Old age does not appear to be directly correlated with tooth loss, as there are some older individuals with all of their teeth and younger individuals that are missing a substantial percentage.

Figure 3.14. A juvenile with a regenerated tail. (Photo from the author's collection)

BLINDNESS

Occasionally a wild alligator will be blind in one eye, and less frequently an alligator will be blind in both eyes. Usually, blindness is due to an injury in which the eye is actually missing (see fig. 3.10). Possible sources of injury are trauma when securing prey, predators, and fights with other alligators. Humans have been known to have blinded alligators during acts of harassment, and in such cases, both eyes are often damaged. In Mississippi and Alabama there have been a growing number of blind alligators that have been discovered in the wild. Suggested causes have been a combination of drought, parasitic infections, and disease, but the known cause is still a mystery. A 2003 Associated Press article placed the number of blind alligators that had been documented at more than 50 in a one-year period in these two states.

POLYDACTYLY AND FUSED TOES

Occasionally alligators possess a different number of toes than the usual five on each front foot and four on each back foot. The first documented case was in 1947 at Sabine National Wildlife Refuge in Louisiana. A small subadult alligator had eight toes on each front foot, eight toes on its left rear foot, and seven toes on its right rear foot. "Doc" Jim Dixon and Mark Staton (1976) published the first case of polydactylism in Texas at Murphree WMA in two juveniles believed to be from the same clutch of eggs. One had eight toes on each front foot and six toes on the left hind foot. The other

animal had eight toes on the right front foot. Amos Cooper of the TPWD Alligator Program captured a two-year-old alligator at Murphree WMA area with seven toes on the left hind foot. A hunter also brought a front foot with eight toes from an alligator harvested from Jefferson County (fig. 3.15). An alligator hatched at Brazos Bend State Park with fused toes.

MANDIBULAR HYPOPLASIA AND OTHER MANDIBULAR PROBLEMS

Murphree WMA has a 6-foot (1.8 m) skull of an alligator taken on a hunt in which the mandible was not fused. I found an entire nest of alligators that had mandibular hypoplasia (lower jaw is too short) where the temperature was too high, at near lethal levels (fig. 3.16; Hayes-Odum and Dixon 1987; Hayes-Odum et al. 1993). The alligator that I kept alive from this clutch still has the condition, but it is less noticeable. I have occasionally seen adult alligators in the wild with slight mandibular hypoplasia, and a couple of other biologists in Texas have noticed such animals as well. EBay has even had a listing of an adult alligator skull with a shortened mandible that was twisted to one side.

LOWER JAW MISSING

The lack of a mandible is an extreme case of mandibular hypoplasia resulting from a congenital defect, or may it also be due to a survivable injury? One of the color plates facing page 251 in *Wildlife Management: Crocodiles and Alligators* shows a hatchling crocodile that never developed a lower jaw (Webb, Manolis, and Whitehead 1987). Amos Cooper of TPWD

Figure 3.15. This foot exhibiting polydactyly was brought in by a hunter in Jefferson County. (Photo courtesy of Amos Cooper, TPWD, and Monique Slaughter, USFWS)

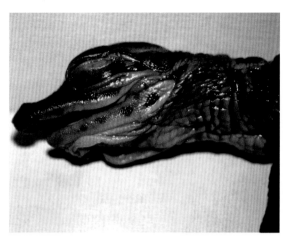

Figure 3.16. A full-term alligator embryo exhibiting mandibular hypoplasia, a shortened lower jaw. This condition affected all of the embryos in the nest and was believed to have been caused by an incubation temperature that was too hot. (Photo from the author's collection)

reports that a hunter brought in a 7- to 8-foot (2.1 to 2.4 m) alligator that had no lower jaw and the tongue was hanging out. It had been taken on private property and was thin. Was this a congenital deformity or a healed injury? In either case, the animal had managed to survive for some time.

CAUDAL SCOLIOSIS

A crooked spine can be from an injury, such as another alligator breaking another's back and/ or tail, or can be due to a congenital anomaly. My experience with the 1988 nest that incubated too hot at Murphree WMA was that a crooked tail, in addition to mandibular hypoplasia, resulted (Hayes-Odum and Dixon 1997; Hayes-Odum et al. 1993). The alligator that I kept alive from this clutch seemed to outgrow the scoliosis. For the first several years, she seemed to be off balance when moving on land, swimming, diving into the water, or procuring prey. Over time, the scoliosis became less noticeable both in appearance and in her movements, until finally it was no longer visible. Jim Tamarack, who worked for the New York Zoological Society on its St. Catherine's Island, Georgia, offshore island sanctuary, reported seeing a subadult alligator in Georgia with an S-curved spine that was unable to swim but lived in the shallows like a turtle.

TUMORS AND WALLED-OFF INFECTIONS

K. J. Lodrigue, who formerly worked with Amos Cooper and Monique Slaughter in TPWD's gator program and then became a TAMU grad student, shared with me that most tumorlike conditions he had seen in gators were really walled-off infections. A tumor is abnormal, living tissue, and a walled-off infection is dead material like pus or dead cellular material that a gator is ridding from its body. An alligator at Brazos Bend that frequented the bridge between Pilant Slough and the south side of Pilant Lake had a rounded structure on top of and near the end of its snout that was pigmented like the rest of its body (fig. 3.17). It

Figure 3.17. An alligator exhibiting a tumor-like growth on its snout that may actually be a walled-off infection. Due to the location, it is possible that there was involvement with a tooth. (Photo by Patricia Rogers)

is possible that it might have had some involvement with a tooth because of its location. No attempt was ever made to capture the alligator and conduct a biopsy of the growth. When I worked at the Houston Zoo in the early 1980s, some of the adult alligators in the alligator pond had "tumors" on the external body surface that may have been partially or all due to walled-off infections.

TEETH GROWING THROUGH THE MAXILLA

Teeth growing "upside down," unicorn-style through the maxilla (upper jaw) have been reported in two crocodilian species. The reason for this abnormal tooth growth is unknown. One was a wild-caught female crocodile in Costa Rica documented with a photograph by Luz Barrentes-Bahder to a crocodilian discussion group in 2010. The other was a captive adult Nile crocodile at Busch Gardens, Tampa, Florida, reported by Michael Malden in response to hearing about the anomaly in the crocodile in Costa Rica. Although no American alligators have been reported with this tooth anomaly, it is mentioned here as something known to occur in other crocodilians.

ALGAE GROWTH ON BODY

The growth of algae or moss on alligators seems to come up repeatedly in various books on alligators, so it is a point worth discussing just to set the record straight rather than to describe a phenomenon that is not commonly observed or is misunderstood. According to Glasgow (1991, 10), it seems that the source of this notion started with a comment by Jean-Bernard Bossu in the mid-1700s that he had seen "crocodiles

so big and so old that they have moss on their heads and backs" (1982, n.d.). Glasgow states that Frank and Ada Graham (1979) found that "'moss' is not a sign of age at all. It is an algae that grows on unhealthy alligators and is not usually seen on healthy animals in the wild." He pointed out that it is also possible that duckweed or some other aquatic plant could have really been the "moss" that Bossu wrote about. Neill (1971, 289) discusses the green alga that sometimes grows on alligators and notes that it is "probably one of the several species that grow more often on turtles" and that "an alga-infested alligator is usually ailing, senile, or penned improperly." He further notes "a green aquatic alga, whether growing on animate or inanimate objects, will flourish best in clear, sunny water."

It is common to see an alligator covered in duckweed at an inland site, such as Brazos Bend, where this tiny but numerous plant covers the water's surface. An alligator swimming in such a water body will surface above the water, or leave the water entirely, covered with the tiny plants and looking like some kind of creature from a low-budget horror "creature feature" (fig. 3.18).

I have seen the water's surface covered in a scum of algae, and it is possible that an alligator could get covered with it in the same manner as duckweed. I have never noticed algae growing on an alligator, but I have seen it on turtles. The filamentous algae do somewhat resemble terrestrial moss plants. No one can be certain whether the moss Bossu referred to was duckweed or algae, but a modern observer of the presence of either should be able to easily

Figure 3.18.
Some of the
references to
moss or algae
on alligators'
bodies may
actually be to
duckweed, a
floating plant
that can cover
an entire pond,
lake, or marsh.

discern which of the two it is and if it is actually growing on the animal.

COLOR ANOMALIES

Texas alligator farmer Jim Broussard has had seven or eight albino alligators hatch out through the years from eggs collected from wild nests. In the late 1990s, he had five albinos in one year. These albinos have pinkish eyes and a "dirty white" skin with the normal yellow striping (fig. 3.19). Albinos lack the dark pigment melanin and get their pinkish eye color from blood vessels that are not obscured by pigment. Albino alligators lacking only melanin tend to be yellowish from the xanthin (yellow) pigments that stand out.

The famous "white" alligators from Louisiana, which have been displayed in a number of zoos in the United States, are leucistic. They have very white skin and blue eyes (fig. 3.20). Leucistic animals have the pigment cells but lack an enzyme needed to activate the pigments. Albinism affects the entire animal, whereas leucism can be incomplete, as evidenced by the one or more dark spots of melanin on at least some of the leucistic alligators discovered as youngsters in the wild in Louisiana. In 2003, three hatchling leucistic alligators were found in South Carolina mingled with a pod of normally colored youngsters. To date, no leucistic alligators are known from Texas. Albino and leucistic alligators have the odds

against them regarding survival in the wild. They are at high risk for predation because of their lack of camouflage and are prone to sunburn because they lack melanin's protection. Therefore, any such animals seen in the wild will most likely be hatchlings.

There can be other color aberrations in alligators that are not quite as striking but stand out. At Brazos Bend in the late 1980s, I noticed a light brown (hypomelanistic) subadult alligator about 3.5 feet (1 m) in length basking near the shore of 40-Acre Lake on several occasions. It was during the drought, and I suspect that another alligator cannibalized it, as it was more conspicuous than normally pigmented individ-

Figure 3.19. A young albino alligator from Texas. (Photo by K. J. and Kasi Lodrigue)

Figure 3.20. Leucistic alligators look white and have blue eyes. Leucistic alligators from Louisiana have been displayed in zoos throughout the United States.

uals. Allen and Neill (1956) and Neill (1971) also mention melanism (deep, uniform black with stripes not showing in juveniles) and erythrism (a cinnamon-red color in the normally black areas) occasionally being seen in alligators.

Longevity

How long do alligators live? Longevity can be viewed in a number of ways. It can be the longest verifiable age in captivity or the wild. It can be the typical age that an adult alligator might be expected to live if it survives to adulthood in the wild. Or we can look at the concept of age-related mortality and calculate the odds of survival on an annual basis.

It is not uncommon for crocodilians in zoos and other captive situations to be alive after 50 to more than 70 years (Snider and Bowler 1992; Weigel 2014). Due to a paucity of records regarding the size and age of many of these long-term captives, it is often unknown whether an animal entered the collection at a very young age, at an older age from another facility with a verifiable age, or as an adult of unknown age. The current record for the American alligator in captivity is 76 years, 10 months as of July 16, 2014, at the Belgrade Zoo, Serbia. The age of this male alligator, named Muja, on arrival at the zoo is unknown. In the United States, an alligator lived for 73 years at the Cincinnati Zoo until its death.

The African slender-snouted crocodile (*Crocodylus cataphractus*) holds the world record for a crocodilian species alive in captivity at over 85 years. A male named Hakuna and a female named Matata arrived at the old Rotterdam Zoo in 1929 and were moved to the new Blijdorp Diergaarde, Rotterdam, in 1943. There are no records on their size or age at the time of their arrival in 1929. The female died on July 19, 2014, and the male was alive as of June 30, 2014.

Adam Britton's website on crocodilians (www.flmnh.ufl.edu/cnhc/cbd-faq-q3.htm) has an excellent discussion of crocodilian longevity. It discusses the problem of whether an animal in the wild or captivity is likely to reach an older age. It notes that the oldest known crocodilian died at the age of 115 years in 1997 at a zoo in Russia (probably just prior to February 16, 1995, as I have a press release of that date announcing the crocodile's death). There is further mention of reasonably good evidence that this animal was captured in the 1890s as a 5- to 10-year-old. Evidence is presented that suggests that the largest species live to be about 70 years on average, with some individuals exceeding 100 years. The method of looking at growth rings in bones and teeth to determine age and their drawbacks is discussed. I will mention for the historical perspective of alligators in Texas that some leg bones were collected by Bruce Thompson, TPWD Alligator Program leader, to ascertain the age of a portion of the adult population through the growth rings. He collected them in 1984, during the first alligator hunting season in Texas. The bones were to be sent to someone in Europe for examination. If the samples were ever sent to be processed, no data resulted.

Natural and Human-Caused Mortality

Natural mortality (i.e., death not caused by humans) includes predation, cannibalism, environmental mortality, and senescence (fig. 3.21).

Predation, being killed by another animal, is an important cause of mortality in young alligators. The smaller an alligator is, the more potential predators it has. Common predators are fish, birds (both aquatic and nonaquatic), and mammals. Florida alligator researcher James Kushlan reported to me that in Florida hawks, gar, and bass are responsible for high predation of young alligators. He further commented that heron and wood stork predation did not occur often except in dry conditions. Actual sightings of alligators being taken by predators are rare. K. J. Lodrigue (former TPWD biologist) saw an actual predation on a juvenile alligator, a 1-foot (0.3 m) animal, which was eaten by a great blue heron in a river during midsummer. Most data are in the form of stomach-content analysis of various predators. For example, hatchling alligators have been found in bullfrog stomachs. Perhaps the smallest predator is the red imported fire ant, which in sufficient numbers can kill young alligators when pipping from eggs or when newly hatched and at the water's edge.

Cannibalism, the predation by an alligator on an alligator smaller than itself, is a cause of death generally restricted to juveniles and subadults. It seems to be most prevalent in the

Figure 3.21. The remains of this alligator that appeared to have died from some type of natural mortality are being scavenged by black vultures.

subadults, especially in low water or other situations in which high densities of alligators occur. The critical period appears to be related to the alligator no longer being under a parent's protection and not having an established territory. Although most cases of cannibalism that I am aware of involve the alligator being eaten, sometimes the alligator is simply killed. Perhaps these instances are motivated by territoriality alone, and the predatory alligator was not hungry at the time. I have been able to predict during drought when (and to some extent where) cannibalism will occur at Brazos Bend. Aside from risk during drought, a population that exceeds its carrying capacity is also at high risk for cannibalism. I have heard of at least two cases in Texas during the alligator hunt when hunters captured 10- to 12-foot (3.0 to 3.6 m) alligators on lines and found that 4-foot (1.2 m) alligators were the ones that swallowed the baited hooks. In one incident, an alligator in coastal marsh near the Murphree WMA was killed by a larger alligator and was found in the larger animal's jaws by the hunter. The other took place at an area adjacent to Brazos Bend State Park and involved a harvested alligator whose stomach was cut open, revealing a smaller alligator (which contained the fishing hook) inside. In both instances, the larger alligators took advantage of the subadults being tethered to a line and in turn went from being predators to prey themselves.

Environmental mortality includes deaths due to freezing weather, hurricanes, drought, flooding, and disease. It also includes rarely known accidents in which an animal falls, is hit by something, or becomes trapped. For example, the one alligator that I am aware of succumbing in such a situation was found dead in Florida on a riverbank stuck between four large rocks. Evidently, the alligator became trapped between the rocks while climbing up the steep bank and could not extricate itself because of the steepness of the bank and its body being positioned so that it could not get its limbs underneath to permit mobility (Pike and Smith 2005). Mortality due to environmental causes can affect any age class. However, as mentioned previously, the subadult size class is often more at risk to the extremes of Mother Nature if they do not possess a quality territory.

Senescence is death from causes related to old age. In the absence of human intervention, an alligator that survives to adulthood and possesses a high-quality territory in prime habitat stands a good chance of living a long life.

Human-caused mortality occurs from hunting and non-harvest-induced deaths. Hunting includes scheduled harvests and poaching and usually involves animals 4 feet (1.2 m) or longer. Non-harvest-induced death consists of any non-harvest death that is caused directly or indirectly by humans. Road-killed alligators or those that drown in fishing nets are some common examples. At Brazos Bend State Park, one adult alligator is known to have drowned by becoming entangled in heavy-gauge fishing line and another by getting stuck trying to climb through the underwater structure beneath a wooden fishing pier. Another example is an alligator that was killed and possibly eaten by another alligator while on a baited line during the harvest season.

Factors Affecting Individual Age-Related Mortality

Age-related mortality is highest in the juvenile and subadult groups, and there is a bottleneck in a population in the transition from the subadult to adult size class (fig. 3.22). The subadults are making a big lifestyle change in that they are no longer under parental care and must find a territory of their own. There is a high degree of various types of mortality in this size class, including cannibalism and environmental factors. If one is trying to construct a predictive model of survival rate of each age/size in a given population, the rates are based on the chance that any alligator at hatching will survive another year or reach the next size (usually based on total body length in foot or meter increments). These rates are really an average of the predicted survival rates of individual animals, which can vary greatly, especially in the first couple years of life.

The odds on survival actually start before hatching. The age, experience, social status, and quality of territory of a given female alligator can have a large influence on how many viable young she will produce and how long they will live. For example, egg viability tends to be lowest in young females just starting to reproduce and in those older females whose reproductive life is near the end. Both young and old females experience higher levels of infertility than those females in their prime. They also may have smaller than usual eggs or irregular eggs in regard to shape or other external appearance such as the eggshell matrix. As in human females, the embryos of older crocodilian females may be subject to a higher percentage of defects and a higher mortality before they reach full-term development.

The experience level of a female can have a profound effect as well. A young female may not be good at constructing a nest or may damage or crush the eggs while she is laying or securing them in the nest (reported to me by TPWD). After the eggs hatch, an inexperienced mother may not know how to protect her offspring from predators and may lose most or all of her young within a few months. An alligator on my land that laid eggs for the first time in 2001 (all four were infertile as no male was present at the time) had a successful hatch of at least 15 offspring of her second clutch in 2005. She swam across the water with babies on her snout, back, and tail, and her other babies were seen close by. Occasionally babies were noticed basking on sandy parts of a nearby island. Suddenly, no babies were seen, and the female became increasingly more camouflaged and did not bask out in the open that fall or the following spring. Presumably she lost some of her young to the numerous resident predators, including bullfrogs, a red-tailed hawk that roosted nearby, herons, and egrets. Consequently, she learned to keep herself and her offspring inconspicuous.

Surprisingly, she reappeared out in the open in April 2007 with at least 13 babies now 20 months old at her second den site at the opposite side of the pond from her original den. Either we underestimated the original number of offspring, or she had a fast learning curve after having a minimal loss of young and was able to save most of them through a change in her behavior. She went to the deeper-water

Figure 3.22.
The subadult
size class has
a relatively high
mortality rate.
No longer under
the protection
of parents, it
is subjected to
cannibalism,
as well as
environmental
mortality,
because it does
not possess
a high-quality
territory.

pond immediately adjacent to her second den site and stayed close to the shore where she could keep watch over her offspring. By the end of April, the juvenile alligators were seen in small groups or singly in various parts of the shallow pond, including the nest site and second den site, and in the adjacent deeper pond in the shallow edges. By May, at least several offspring had left both ponds and were seen in drainage ditches some distance from the pond and in grassy areas halfway between the shallow pond and an adjacent bayou.

A typical loss of young for a mother might be about 50% within the first year. I arrived at this estimate from my mark-recapture study and by having watched several alligators at Brazos Bend State Park whose young could be viewed and counted over time. After hatching, the number of offspring that were counted on each visit gradually dwindled. By the following spring when they came out to bask, about half as many were counted as the greatest number seen after hatching.

As with human mothers, some alligator mothers are more competent than others and have a greater survivorship of offspring. For a number of years at Brazos Bend State Park, we watched a female we called "Big Mama" that was over 8 feet (2.7 m) in length and sometimes had up to three different age classes of juveniles with her. If an alligator nest had to be destroyed because it was on a major hiking trail and the eggs incubated artificially, it was a common practice to release the babies with her after they hatched to give them the best chance for survival.

The length of time that offspring spend with a female varies. A youngster approaching three years of age is about the upper age level seen in association with a female. Are these older youngsters all that survived of their pod, or do some just leave sooner than others? That is a question I cannot currently answer but wonder about. The distance that a juvenile/subadult travels after leaving the protection of a parent is another unknown. We find older juveniles and subadults traveling in areas that would suggest they are in transit and looking for a place to set up a territory. It is difficult to know where these animals originally came from and how far they traveled. For example, near the entrance of Brazos Bend State Park where the main office is located, water seasonally sits in an area around a mature tree. One year an older juvenile and young subadult "hung out" in this area. The closest known "alligator nursery" that these individuals could have come from was the prairie potholes some distance away. It seems as though the common time for these migrations occurs when the weather is not overly warm or cold and wet conditions exist so that trips overland would coincide with the presence of standing water, which could help prevent dehydration and assist in thermoregulation.

Sometimes females may take their offspring to an area and then leave them. This was observed with a female at Brazos Bend who took her yearling offspring from Pilant Slough to nearby Elm Lake, stayed with them four or five days, and then left them on their own. They dispersed from the area within a week. This is somewhat similar to what I previously described for the female that nested on my property in 2005. In 1999, one of the alligators that I captured and tagged as a juvenile was recaptured at Elm Lake near its original capture site nearly 15 years later. Evidently it was able to stay in the same area through adulthood without a problem. It was sexed as a female, and that may have contributed to its not leaving. Is this an example of a category of parental behavior known as social assistance, whereby offspring develop home ranges that overlap with that of their parents? Additionally, it may not have moved because Elm Lake is prime alligator habitat and large enough to support numerous alligators.

It also had to exist in a situation where the population of the lake was not exceeding carrying capacity, or this species of top predator would have probably "self-regulated" by cannibalism just as in drought when available space shrinks. I wonder in such a situation how it interacts with its parents after it reaches subadulthood and adulthood. Is survival of a subadult enhanced if it is able to remain in or near the area in which it spent its early life? In general, the larger an alligator grows, the fewer predators it has. For a juvenile that leaves the care of its parent, it seems likely that its predation rate is decreasing due to its growth but that its risk of cannibalism is increasing due to lack of parental protection, relatively small size, lack of its own quality territory, and risk that it will cross into another alligator's territory. Are some large juvenile and subadult alligators at a greater risk than others because they have to migrate? The answer could provide better predictability to alligator population mod-

els, given that the greatest mortality is in the juvenile and subadult groups.

Reproduction

Reproduction is seasonal and takes place only once per year. Not all sexually mature alligators participate in reproduction in a given year (including those that have reproduced previously), and reproducing females have a maximum of one clutch of eggs per year. The mating period is characterized by a lot of movements, grouping behavior, vocalizations, and male/male and male/female interactions. Instances of male/male combat during territorial disputes sometimes occur.

Participation in Reproduction

Participation in reproductive activities does not always coincide with the onset of sexual maturity. The animals that participate in reproduction must also have well-established territories. McIlhenny (1935) noted that 9 years, 10 months of age was the first time of nesting for a female that he had marked as a hatchling at Avery Island. A female that I kept captive in an outdoor enclosure first laid eggs when she was nearly 13 years of age. Based on a number of crocodilian species kept in zoos, breeding seems to typically occur in crocodilians at about 10 years of age or older.

Although both sexes may breed for the first time at about the same age, there is a difference in their growth rates and the onset of sexual maturity. In the first years of life, both sexes of alligators grow approximately 1 foot (0.3 m) per year. Then the female's growth rate slows down relative to the male's. Both reach sexual maturity when they are at a length of approximately 6 feet (1.8 m). A sexually mature male may be 6 or 7 years of age yet does not have the social status to breed for several more years. Meanwhile, the slower-growing female has the time to secure a territory of her own so that she is more apt to reproduce at the onset of sexual maturity. Alligators in the northernmost portions of their range tend to have a slower growth rate and reach sexual maturity at a later age than those ranging farther south.

Timing of Reproduction

Alligators have a shorter and more discrete period of courtship, mating, and nesting than other crocodilian species that are not constrained by living in a temperate climate. Basking and feeding get under way in late winter/early spring, followed by courtship and mating, as well as territorial disputes. Bellowing activity, a reflection of increased social behavior, typically starts in March and remains high in April. Throughout May, courtship and mating are at a peak. June typically is the nesting month, with three to four weeks elapsing between ovulation and nesting. The earliest that I know eggs have been laid in Texas was May 26 at Brazos Bend, with the latest being in early July. Approximately mid-June is the peak of egg laying for most years.

Temperature and photoperiod (day length) vary seasonally and are two major cues for the timing of reproduction. In Louisiana, Ted Joanen and Larry McNease (1979, 1989) found that March–May ambient temperatures were associated with variability of the timing of nesting, with the earliest nesting occurring when the March–May ambient temperatures were the highest.

As noted earlier, environmental stressors such as drought and flooding can have a negative impact on reproduction. Sometimes the timing of this impact seems to be different than might be expected. Researcher Louis Guillette (pers. comm.) studied plasma levels of estradiol (female estrogen) and vitellogenin (protein precursor to egg yolk) in female alligators and found that they begin to prepare for reproduction in the fall rather than spring. These data provided an explanation for observations I made on alligator reproduction at Brazos Bend during a drought that began in spring 1988 and ended in spring 1989. Since the drought had just started at the beginning of the breeding season in 1988, the amount of bellowing and mating activity did not appear to change from that observed in previous years. However, in spring 1989 reproduction was adversely affected even though the drought had ended by the onset of the spring mating season. Bellowing and mating behavior were depressed, and nests were present only at the perimeters of lakes that retained substantial areas of permanent water during the drought. Thus, the critical time for the onset of female reproduction for the 1989 nesting season was in fall 1988 when the alligators were still subjected to the drought (Hayes-Odum and Jones 1993).

The awareness of the timing of this lengthy process of preparing for reproduction in various crocodilian species is now being used as a management tool in captive populations to avoid subjecting the animals to stress during critical periods when reproduction could be adversely affected. For example, moving crocodilians to a new exhibit or shipping them from one zoo to another involves not only choosing a time of year that is not extremely hot or cold but also one that does not adversely affect their reproduction in the coming year. This strategy of minimizing stress during periods critical to reproductive success also has management implications for crocodilians in the wild on lands where human-made changes in habitat or water level are made.

It appears that environmental stressors sometimes may lead to an unusual sequencing and alteration of reproductive events. At Brazos Bend there were two peaks of breeding-related behaviors in 1992, and nesting coincided with the second peak (Hayes-Odum et al. 1996). The first peak of courtship and mating was on May 11, with the first bellow having been heard six weeks earlier. All mating behavior declined after mid-May, and the first nest was discovered on June 17. From June 18 to June 21 there was an intense, sudden onset of bellowing, grouping behavior (especially in pairs), and possible copulation that stopped as abruptly as it had started. Four of the 21 nests discovered at Brazos Bend during 1992 were fully formed rather than false nests, which are incomplete, but did not contain any eggs. Three of the nests without eggs were actively defended. The other nest without eggs was guarded for approximately a day before the female appeared to lose interest in it and did not chase human intruders away. The dates of nest construction were known to be June 17, 23, and 30 for three of the nests without eggs. The earliest known nests containing eggs were completed June 30–July 1.

These events were believed to be related to unusual flooding that occurred from the end of December 1991 through mid-March 1992. Separate water bodies were joined by the flooding, resulting in a single large body of water over much of the park with water levels 4 to 5 feet (1.2 to 1.5 m) above normal in some areas. Furthermore, there did not seem to be any unusual patterns of daily or monthly temperatures for March–July when compared to long-term data, so it was thought that the unseasonable flooding was likely the proximate cause. Fully formed nests without eggs were also reported by Franklin Percival (pers. comm.) at Lake George in north-central Florida in 1990. In this instance, nesting took place one week earlier than usual, but there were no abnormal water levels or other unusual conditions that might adversely affect nesting.

Mating

During the peak basking season in fall and spring, adult alligators often bask in groups without any aggression or seeming concern of territorial boundaries. As soon as they begin feeding and the breeding season is ready to start, an entirely different picture emerges. The alligators move considerable distances, groups of adults congregate, and there is a big emphasis on territoriality. Do alligators have a set territory, or is it a seasonal occurrence for males to try to claim and defend territory? Although territories (at least home ranges) are maintained year-round by both males and females, the breeding season involves a very conspicuous and heightened territoriality by males to gain access to mates and limit access of subdominant males to mates. Territory borders are patrolled by dominant males. At close range, dominant males try to make themselves look larger and more formidable by exposing as much of the top of their head, back, and tail as possible; subordinate animals expose only their heads. Bellowing displays and head slaps convey the presence of dominant males from a distance. Both audio and visual displays can repel less dominant males and attract females. A dominant male may interrupt the courtship of a subordinate male and drive it away by chasing it and making open-mouth lunges. It is not uncommon for the dominant male to breed with the female afterward. When combat occurs between males, head ramming and actual biting down takes place. Generally the limbs and the base of the tail are the portions of the body that the opponents grab with their mouths. Even the dorsal surface of the body, which is covered in bony plates, can be injured and bleed (see fig. 3.13). The fight stops when one individual backs down.

Females also bellow. Each alligator has a distinctive voice, which can be recognized by other alligators that reside in the general vicinity. When one alligator begins to bellow, others often join in until a chorus of alligators ensues. I might add that a noisy boat engine or a truck backfiring can often elicit a bellow from a nearby alligator as well. It has been postulated that adult alligators tend to be noisier than most other crocodilian species because much of their habitat is highly vegetated, which results in poor visibility over a distance. By both sexes bellowing and all animals having unique voices, each animal in that chorus knows the location of specific individuals in the chorus and can make a decision on what its next move might be.

After the attraction phase of courtship has occurred, pair formation begins. Females move to males or males to females, but it often is the female that initiates courtship. I noted a female at Brazos Bend having a male come to breed with her in an area of Pilant Slough where she had been seen with the previous year's offspring. Several days later she walked quite a distance on the hiking trail and crossed over to Pilant Lake, where she bred with another male. Thus, a female may not be consistent in whether she approaches a male or has a male approach her for courtship and breeding.

Mating takes place in open water, which is deep enough to facilitate the process of copulation. Mating requires proof that the other partner is receptive and can be trusted, since either the male or female would be capable of seriously hurting a potential mate. Crocodilians in captivity have been known to kill potential mates that have been placed with them, even in cases where considerable time and care has been taken in their introduction. Snout lifting, which exposes the throat area that is unarmored and thus vulnerable, is a principal signal that the other alligator is willing to engage in courtship (fig. 3.23).

Precopulatory behavior involves a lot of touching of the heads and snouts, nudges, crossing snouts over each other, and blowing bubbles (fig. 3.24). The male sidles alongside the female, and they swim together in the

water. Stroking, riding on top of each other, and pushing each other under the water are other activities seen prior to actual copulation.

Copulation is hard to see as it takes place underwater and may last 5 to 15 minutes. During copulation the male dorsally mounts the female and positions his tail and vent beneath hers so he can insert his penis into the female's cloaca. Both male and female possess a cloaca, which is a common chamber into which digestive, urinary, and reproductive ducts open.

An instance did occur at Brazos Bend in which the male forced a female to mate with him (Hayes-Odum and Jones 1993). A large female was cannibalizing a subadult alligator that swam into her territory, and a male chased and mated with her while she ate it. Copulation lasted 5 minutes, with the female completely submerged and exhibiting signs of stress. She shook her prey violently, growled at the male, and crawled out of the water onto the bank to continue eating. The next day the male picked the posterior portion of the carcass that was left and began eating it while the female was out of the vicinity. It appears that this female was more interested in protecting her prey than in

Figure 3.23. This female lifts her snout to signal to the male that she is ready to participate in courtship.

Figure 3.24. Interactions involving the heads of alligators during courtship include touching of the heads and snouts, nudges, crossing snouts over each other, and blowing bubbles.

breeding. The male was able to breed with her without fear of attack, as she would not let go of her prey, probably because she worried that the male would steal it from her. It is obvious that the male did have an interest in the prey, as he took it the next day.

Alligators are generally considered to be polygynous, in that the male usually breeds with multiple females. Although the territory of the male may touch on several female territories, and he may breed with several females, a female may breed with more than one male as well. Thus, a female may have her clutch of eggs fathered by more than one male in a given year. A study in Louisiana at the Rockefeller Refuge (Davis et al. 2001) found that of 22 nests, offspring of 15 nests were sired by a single male, offspring of 6 nests were sired by two males, and offspring of 1 nest were sired by three males. The same male and female may also breed with each other repeatedly during the same breeding season.

Nesting

After a peak of courtship and mating, bellowing and mating activities decrease and the groups of alligators disperse (see Hayes 1992; Hayes-Odum et al. 1993). Females that will produce eggs begin to prepare for nesting. There are three to four weeks between ovulation/fertilization and egg laying. The female chooses a site on which to construct her nest. Prime nesting sites are usually located near the den and associated "hole" that she has excavated. When I have seen nests some distance away from the den area, the site seems to be prone to flooding, so the nest is located at a higher elevation. It may be a location where there is a steep slope, and the nest is built at the top of the slope. Sometimes the water body is a shallow basin with a very gradual slope upward, and the alligator moves as far away from the water body proper as she feels necessary to avoid flooding. It is not uncommon in these cases for the nest to be built on a pile of logs if there are some in the vicinity. It also seems that the alligator is diligent about making sure that within the nest the egg chamber is above the flood level in general (fig. 3.25).

With abnormally high water levels at Brazos Bend during nesting season, we noted that nests were built in places where they normally would not be (e.g., adjacent to hiking trails). However, changes in nest location are certainly not predictive of future rises in water

Figure 3.25. Female alligators seem to have the ability to build their nest so that the nest cavity is above the waterline. However, sometimes nests may be flooded. The embryos perish if exposed to water for a sufficient length of time. Usually, this flooding is associated with rain from hurricanes. Otherwise, in years where water levels are higher than usual at the time of nesting, the females will build their nests on higher ground.

level during the nesting season, as evidenced by nests that flooded at Brazos Bend and elsewhere. Nests are sometimes constructed around the trunks of bushes or very small trees. This may make nest construction easier and hold the nest together more effectively. Perhaps they may also provide some camouflage from potential nest predators.

Nest Construction

Nest construction may take from several days to well over a week from the start of the construction to the laying of the eggs. The longest that I have seen a nest under construction is 10 days. Sometimes one or more incomplete "false" or "test" nests are created before a nest is completed that will hold the eggs. It is thought that females find these unfinished nests, or the exact sites where they are located, to be unsatisfactory in some way, so they abandon them before completion. Another suggested reason is that they are constructed to confuse predators that may leave the real nest alone after finding no eggs in the "false" nest. It is not uncommon for a female to build her nest in the same spot for a number of years, sometimes on top of the remnants of a nest from a previous season.

The first step in nest construction is site preparation. The female clears the area where the nest will be constructed. She also creates one to two trails leading from the water to the nest, which remain trodden down from her trips back and forth to the water during nest construction and then from visits to the nest during the incubation of the eggs. The arduous process of nest building is very time consuming

and requires experience rather than instinct alone. McIlhenny (1935) described in detail how a female constructed her nest over a three-day period as he watched from a nearby blind. The following is a condensed version of the process he witnessed.

The first day the alligator began by biting off vegetation with her mouth and creating a pile of this vegetation near the center of the resulting clearing. She repeatedly went to the perimeter of the clearing that she was creating and brought back a load of vegetation. Her method of biting off rooted vegetation was to grab a mouthful and pull backward like a tug-of-war in the direction of the pile of nest material. The weaker stems broke off, some plants came up by the roots, and the strongest vegetation was broken off by violent shakes of the head. She even bit off several trees that had trunks up to 3 inches (7.6 cm) in diameter and placed them across the future nest. She laid her body across the tree trunks to hold them in place and tore off their limbs by crushing them in her mouth and shaking her head violently to break them off. The limbs were utilized in the nest, and the trunks were carried to the side of the clearing and discarded. The result was a nearly round pile of stems, twigs, and leaves approximately 6 feet (1.8 m) in diameter and 18 inches (45.7 cm) in height.

The female then repeatedly crawled over the mound and worked the material from the edge to the center until the structure was well packed and flat on top. The female had spent most of the day working on the nest, and when McIlhenny returned the next morning, he discovered that she had also worked at night. The clearing had increased in size, and the resulting vegetation was piled loosely around the base of the developing nest. So the female had spent nearly an entire day and a portion of the night for the first day of nest building.

The alligator arrived at the nest early the next morning and carefully surveyed the nest mound while crawling around its base. She then crawled on top of the nest pile with her tail on one side and her head on the other. She reached down and grabbed the loose and projecting material from the mound base and held it firmly in her mouth; she then backed across the nest, bringing the material across the center to the opposite side. It was deposited and then firmed down by the weight of the female's body pushing forward across to the edge again. This process continued over two hours until two complete circles had been made of the nest, and it was nearly a foot higher. Then she crawled down and looked at it all over the sides and the top, but paid the most attention to the top. She then went to the top of the mound and placed her hind feet near the center. Lying flat, she braced her front feet into the outside edge of the nest material. Then she drew up and pushed back her hind feet slowly, one at a time, turning slowly around the rim of the mound. After she had made a couple of circles, the result was a center that was hollow with the outside edges raised about a foot above the center. This process took nearly three hours.

The female then went to the water and pulled up wet and muddy rushes by the roots, along with the accompanying mud, and made nine trips of huge mouthfuls of rushes, which she put in the nest hollow until it was much higher than the sides. This took her over three hours. She again surveyed the appearance of her developing nest, went back to the top, scooped material from the top, and placed it on other parts of the top until it was approximately 3 feet high (0.9 m) and looked like a haystack.

The third day, when McIlhenny returned to the nest, it appeared that the female had not done any work during the second night. Shortly before noon the female came out of the water to the nest and climbed on top. She proceeded to hollow out the center of the nest with her hind feet by drawing forward and then pushing back and down with force while turning her body around the nest rim as the hollow progressed. McIlhenny commented that this hollow was not as wide as the one that she previously made but was deeper and made much more quickly because of the softer vegetation involved.

The female began laying eggs shortly after noon. She placed her hind legs on each side of the hollow and laid most of them at a rate of one egg per nine seconds. After approximately 30 eggs had been laid, there was a pause of 16 minutes after which she laid six eggs. Then she had a pause of 7 minutes and laid five more eggs. After resting for 8 minutes, she pushed her body to one side of the nest top, took a large mouthful of rushes and other nesting material from the side rim, and dropped it on top of the eggs. She continued this process until the cavity was filled to the top. Then she went to the pond to get more rushes, dropped them on top of the nest, and used the weight of her body to press the wet material into the nest.

Figure 3.26. A common reed (*Phragmites australis*) nest often found in Texas coastal habitat. (Photo from the author's collection)

Figure 3.27. A marshhay cordgrass (*Spartina patens*) nest that is associated with Texas coastal habitat. (Photo from the author's collection)

She made six more trips to get rushes, mud, and partially decayed matter from the water. Then she crawled along the slope of the cone instead of the very top, after the center fill got higher, to slick down the nest surface. After finishing her construction shortly after 4:00 p.m., she walked around the nest to inspect it before returning to the water.

Types of Nests

The type of vegetation found at coastal and inland sites in Texas tends to vary, as does tree cover. A coastal area such as Murphree WMA has an aerial nest survey flown each year by TPWD, as one type of census to keep tabs on the alligator population. However, it is impossible for TPWD to conduct aerial surveys for alligator nests at Brazos Bend, as the majority of nests are under the cover of trees and are not visible from the air. Prairie nests would show up, and some areas of the park that are now covered by the invasive introduced Chinese tallow trees (*Sapium sebiferum*) would have been visible by air in the past. So inland habitat itself is not the deciding element on whether or not nests are visible from the air.

When I was conducting the nest study portion of my dissertation research at Murphree WMA (coastal) and Brazos Bend (inland), I selected nests based on the types of nesting materials present at each site. The nests at Murphree were typically constructed of common reed (*Phragmites australis*) or marshhay cordgrass (*Spartina patens*). Those at Brazos Bend were usually made of soil or giant cutgrass (*Zizaniopsis miliacea*). As previously mentioned, almost all nests in a given year at Brazos Bend are under tree cover. Common reed is actually a tall grass that grows up to 16 feet (4.8 m) in height and has thick, woody stems. It occurs in large stands and grows in water up to several feet deep or along moist shorelines. Nests made of common reed were usually located on levees and contained a substantial amount of soil with a high humus content. These nests often contained fire ants. Common reed nests were usually surrounded and shaded by the live common reed plants, which grew close to the nest (fig. 3.26).

Marshhay cordgrass is a very fine grass, reaching a height of up to 5 feet (1.5 m). It grows in small clusters or dense tufts in coastal areas, usually in the tidewater zone. The cordgrass nests were larger in size than common reed nests, had little to no soil, and lacked fire ants. They looked like haystacks. Unlike the common reed nests, the cordgrass nests were not shaded from above (fig. 3.27).

The most common nest type at Brazos Bend in a given year has been soil nests (79 % in 1990). They contained loose earth, minimal vegetation, and often pieces of sticks (fig. 3.28). Fire ants have been frequently encountered in these nests.

Giant cutgrass reaches a height of up to 10 feet (3 m) and grows in water or next to water. It gets its common name from the edges of the leaf blades, which are very sharp and cause cuts to the skin on contact. The cutgrass nests were in the form of an aggregation of short pieces of stems like the common reed nests but lacked any soil (fig. 3.29).

My later work at Brazos Bend with alliga-

Figure 3.28. Nests of soil are typically found in inland habitat where tree cover occurs rather than in prairie or open marsh habitat. (Photo from the author's collection)

Figure 3.29. Nests of giant cutgrass (*Zizaniopsis miliacea*) are associated with inland habitat. (Photo from the author's collection)

tor nesting documented the existence of grass nests resembling the cordgrass nests in appearance and size. They were composed of Florida paspalum/beadgrass (*Paspalum floridanum*) or Johnsongrass (*Sorghum halepense*). Through the years at Brazos Bend, I have found nests from time to time that used unusual building components or were constructed in an unusual way. I have found nests composed of aquatic vegetation such as bulrush, cattails, water lilies, and lotus. Although leaves are sometimes added to soil nests, they can also be the major nest component. One nest was constructed entirely of decomposed egret guano on an island that had a substrate that was very spongy rather than being made of hard clay as a result of high usage by egrets leaving guano to accumulate and decompose.

There are three nests at Brazos Bend that are memorable as being the most unusual. In the same year at Brazos Bend, two unique nests

were found on which we published an article in *Herpetological Review* (Hayes-Odum et al. 1994). One was built on an island of logs totally surrounded by water (fig. 3.30). Nesting material consisted of grass, leaves, and soil.

The other possessed an egg cavity that was completely below ground level (fig. 3.31). The nest measured 12.5 inches (31.8 cm) at its highest point, and the uppermost egg was 3 inches (7.6 cm) below ground level. The nest was composed primarily of soil mixed with some sticks and finer plant material. We documented four nests that were lower in height, yet none had eggs in a subterranean cavity. Alligators are considered to be mound nesters, and this was the first published account of an alligator building a hole nest.

The third unusual nest at Brazos Bend was at Hale Lake, which is not a high-activity area for alligator nesting or alligators in general. The nest was made of greenbrier (*Smilax rotundifolia*) and grass, and a thin, loose cap formed by the greenbrier was covering the eggs. The eggs were fertile, as we could see banding on them. The approximately 7-foot (2.1 m) female was guarding the nest and lunged out of the water at us but did not approach the nest as we opened it. When we went back to check on the nest, we discovered that someone mowing the area did not realize that it was a nest and mowed right over it, crushing the eggs. There was no moist egg chamber or any real way to protect the eggs from changing temperatures. Had the nest not been destroyed, there was a good chance that it would not have had a successful hatch. The female was probably nesting for the first time, as a nest had never been

Figure 3.30. An unusual nest of logs in that it is in the middle of a marsh rather than on land. (Photo by Clayton McKee)

constructed at that particular location in the past and there appeared to be only a few eggs present. Also, in future years, conventional nests were constructed at the same site, and the female guarded the nests and had successful hatches. A captive alligator, nesting for the first time, laid her few eggs in a cement block. Possibly a lack of nesting material and a desire by the female to hide or cover the eggs explains this interesting "nest." The eggs were not fertile, as no male was present.

The nest that McIlhenny watched being constructed was 86 inches (218.4 cm) across the base and 40 inches (101.6 cm) high. At the Murphree WMA in 1987, the nests chosen for my study were typical for size as well as other aspects. The cordgrass nests were larger than the common reed nests. One cordgrass nest measured 64.2 inches (163 cm) in diameter and 20.9 inches (53 cm) in height, and the other measured 65.0 inches (165 cm) in diameter and 22.1 inches (56 cm) in height. One common reed nest had a diameter of 48.1 inches (122 cm) and height of 16.2 inches (41 cm), and the other had a diameter of 50.0 inches (127 cm) and height of 18.1 inches (46 cm). In 1991, when I was conducting a study of nests at Brazos Bend, 18 nests composed of dirt and stick-type nesting material ranged from 24 to 83 inches (61.0 to 210.6 cm) in width by 36 to 84 inches (91.4 to 213.4 cm) in length by 12 to 30 inches (30.5 to 76.2 cm) in height and obviously showed quite a range in size. A prairie nest at Brazos Bend in 2000, made entirely of aquatic vegetation, measured 85 × 94 × 24 inches (215.9 cm × 238.8 cm × 61.0 cm).

Dennis Jones noted at Brazos Bend that nests were becoming smaller as a result of more trees and fewer open areas (and less vegetation available for nest construction). Alligators use whatever vegetation is in the area to build their nests, and I think that the predominance of nests consisting largely of soil at Brazos Bend is due to the lack of suitable vegetation in these overgrown areas. I noted the problem of less open space about 15 years earlier and suggested that areas, especially islands where a lot of nesting occurs, be "opened up" so that both alligators and aquatic birds have more areas available for their use. Dennis addressed this problem to ensure that quality habitat exists for these vertebrate species. It should be interesting to look at past nest sizes before and after intervention to open up areas to see if the nest size increases substantially.

Figure 3.31. Alligators are known to be mound nesters, and this nest from Brazos Bend is the first on record to have a below-ground nest cavity. (Photo by Clayton McKee)

Eggs, Development, and Hatchlings

When I conducted research in the late 1980s (Hayes 1992), I found that the nine nests from which I removed the eggs to determine the sex of the offspring contained from 17 to 48 eggs. The first year, the minimum number of eggs was 23 and the maximum was 48. The second year, the minimum was 17 and the maximum was 48. Both years separately and together had a mean of 36 eggs. The number of eggs for the nine nests was 17, 23, 28, 29, 41, 45, 46, 48 (two nests). The lowest number of eggs laid that I have witnessed is 4. This very low number was laid by a first-year nesting female in a seminaturalistic outside enclosure where no male was present to fertilize the eggs. Another nest with a low number of eggs laid by a probable first-year nesting female was at Brazos Bend. Both are the nests of unusual composition mentioned earlier. The literature generally lists the range of the number of eggs at approximately 20 to 70 for alligators, with 28 to 52 as the usual number. From my data and those in previous studies, the most common number of eggs for a mature female is between 40 and 50.

It is not uncommon to find turtle or snake eggs in alligator nests. I have discovered turtle eggs in nests that, based on their size and shape, probably were red-eared slider (*Trachemys scripta elegans*) eggs. In one nest at Brazos Bend I found snake eggs, which most likely were laid by the commonly seen mud snake (*Farancia abacura*) or Texas rat snake (*Elaphe obsoleta lindheimeri*), since the other frequently encountered snakes near the water, crayfish snakes (*Regina* sp.), water snakes (*Nerodia* sp.), and water moccasins (cottonmouths)/copperheads (*Agkistrodon* sp.), give birth to living young. It would seem that by laying eggs in a nest that is guarded, these other reptiles would be afforded extra protection from nest predators. Additionally, it seems that if a turtle or lizard nest was found in the alligator nest, the presence of the alligator eggs might deflect the predator and afford these eggs some protection as well. Also, turtles and snakes laying their eggs in alligator nests may be just looking for a suitable nesting substrate, and the protection afforded by an alligator is just a fringe benefit.

Egg Characteristics

The shell of an alligator egg is calcified like a bird egg rather than the leathery composition of a snake egg. The elliptical eggs are white (fig. 3.32). One end is not larger than the other, as in most bird eggs. The dimensions of a typical egg are 1.75 inches wide × 3 inches long (4.45 cm × 7.62 cm).

I weighed the eggs that I removed from alligator nests in 1987 and 1988. The mean weights per clutch ranged from 2.05 to 2.58 ounces (58.5 to 73.7 g), which is comparable to the weight of chicken eggs. When correlating the egg weight with viability (living embryos), viable egg weights were higher than the clutch average for four nests, equal to the clutch average for two nests, and lower than the clutch average for three nests. The most variation tended to be among rotten eggs, which varied from being extremely light (if open and dehydrated) to heavy (if fully hydrated).

Egg viability was separated into the following classes in my study involving nine nests:

Figure 3.32. The egg chamber in this alligator nest has been opened, revealing the top layer of elliptical eggs.

viable, infertile, early dead, late dead, full-term dead, rotten. Viable eggs are those that will hatch living offspring, unless there is a mishap at the time of hatching, such as fire ants biting the hatching alligators, especially if a parent is not available to aid in the hatching. If an embryo dies extremely early, it might be mistaken as an infertile egg because it does not begin to decompose. All of the nests contained viable eggs, ranging from 23% to 98%. Most fertile eggs that die do so early in the incubation period, probably due to environmental extremes or genetic abnormalities. Six of the nests contained early dead embryos in the following percentages: 71, 40, 14, 7 (2), and 2.

Late dead embryos were found in three nests (11%, 9%, 4%). Only one nest had dead full-term embryos (8%), and this was the one previously mentioned in which all viable embryos had apparent heat-induced abnormalities in the form of mandibular hypoplasia (short lower jaw) and caudal scoliosis (kinked tail). If a dead embryo is not too badly decomposed, it is possible to ascertain the stage of development that it had attained before death. Six nests contained rotten eggs in percentages ranging from 3 to 30. Rotten eggs appeared to be eggs in which the shell had been broken or the embryo had died and reached an advanced stage of decomposition. Some or all of the rotten eggs with the broken shells may have been the result of the eggs bursting from the decomposition process.

When the eggs are first laid, there are no external differences between fertile and infertile eggs. However, approximately a day after having been laid, a spot forms beneath the shell at the midline of the egg and is visible to the naked eye from outside the shell. Within a few days, the spot grows into a band that extends around the entire circumference of the egg. As incubation progresses, the band widens, until it finally covers the entire egg. The egg, which started out as being somewhat translucent in unbanded areas, is now totally opaque. This banding occurs as a result of structural changes in the shell due to loss of water from inside the egg and from calcium being removed from the shell as it is used by the embryo. The opaque areas allow gas exchange to take place more easily between the developing embryo and the external environment. The embryonic disc (the tissue from which the embryo and extraembryonic membranes develop) attaches to the shell at the spot that first became opaque. The rate of the spread of the opaque banding over the egg and the growth of the embryo are highly correlated. By examining band width, it is possible to ascertain the stage of development of the embryo and the approximate number of days into the incubation period. In the case of complete banding, incubation can be assigned to a minimum age and stage (45 + days of incubation and stage 24 of Ferguson's [1985] embryonic development scheme).

Although most eggs that were not viable looked indistinguishable from viable eggs (except for rotten eggs), two nests had eggs that were not viable, which had external differences from viable eggs. One was from the nest that had the fewest eggs (17), the lowest viability (23 %), and the greatest percentage of early dead embryos (71 %). All of the eggs were visibly small, and one egg was even smaller than the rest (infertile), one egg had a very oblong shape (early dead), two eggs had calcium deposits on the shell (early dead), and one egg had both an oblong shape and calcium deposits (early dead). The other nest, which had the embryos with deformities, had one egg with a concretion on the end of the shell and the other egg flat on one side (both rotten).

Arrangement of Eggs in Nests
I photographed, as I numbered eggs in a clockwise fashion and then removed them from a nest at Brazos Bend, the pattern from top to bottom. From this nest and the others that I have investigated, there were a few eggs at the bottom and the top, with the bulk being at the center. A map of this nest was made from the photographs, going from top to bottom (fig. 3.33).

Length of Incubation
The higher the temperature is, the faster the rate of development and the shorter the incubation time. The incubation period for alligators is typically 65 to 70 days, with the length of the incubation being correlated with the temperature. However, McIlhenny (1935) did monitor a nest on Avery Island that required more than 100 days for the embryos to reach full term. He attributed it to the nest being made out of soil and "fairly dry," lacking green nest material that would produce heat as it decomposed.

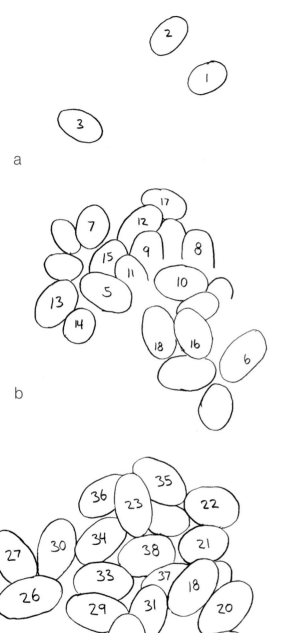

a

b

c

In 1990 at Brazos Bend, the incubation period was greater than 75 days for at least the three nests for which approximate incubation starting dates were noted. One of the nests from 1990 was opened prematurely by a parent or parents after approximately 75 days of incubation. The other two nests hatched at approximately 76 and 78 days. The nest that was opened prematurely was made of soil and was fairly dry inside, as was the nest McIlhenny had noted. However, the other two nests were both composed of grass, so that explanation is not relevant to them. Two other possible explanations of the observed extended incubation relate to monthly air temperatures during the incubation period. The July 1990 mean temperature was lower than that of the 30-year average. Also, nesting was a month earlier than usual; thus, air temperature was lower when these nests were built in late May–early June than it would have been if the nests had been built later in June or early July.

Fluctuations in the timing of nesting/egg deposition have been attributed to a direct correlation with the March-April-May ambient temperatures (Joanen and McNease 1979, 1989). When nesting took place early, temperatures for those months were correspondingly higher. In a comparison of 1990 mean monthly temperatures for those months with the 30-year average (Houston Area National Weather Service data), it was found that tem-

Figure 3.33. (a–d) A nest map made from a series of photographs of a typical nest egg chamber. A clutch has a global shape with fewer eggs at the top and bottom, and most of the eggs in the center. (Tracings made by author from photographs)

peratures were higher for each of the three months of 1990. The previous winter temperatures were also colder than usual; December 23, 1989, was the coldest day with a low of 6.9°F (−13.9°C).

Environmental/Temperature
Sex Determination

In genetic sex determination (GSD), the sex of an individual is decided at the moment of conception by a pair of sex chromosomes (a single one contributed by each parent and carried in the egg and sperm). However, in some animal species that reproduce sexually and have eggs and sperm, there are no sex chromosomes. The determination of sex occurs sometime later during the embryonic development and is influenced by an environmental factor. This phenomenon is termed environmental sex determination (ESD).

In crocodilians, the sex is largely determined by the temperature at which they are incubated during a critical time period when the gonads (sex organs) differentiate into male or female (testes or ovaries). This type of ESD is termed TSD (temperature sex determination). In vertebrates, TSD has been found in some fish species and in many reptiles. It is not known in amphibians, birds, or mammals. In reptiles, the snakes are the only group in which TSD does not exist. The tuataras and all crocodilian species studied thus far have TSD. Many turtle species (including all of the sea turtles) and a few lizard species display TSD (see Valenzuela and Lance 2004).

The critical time period for the sex to be determined by temperature is termed the ther-

mosensitive period and typically takes place during the first third of the incubation period in alligators (days 7 to 21). The timing and length of the thermosensitive period are correlated with the rate of development, which as previously mentioned, is related to the incubation temperature.

Biologists Mark Ferguson and Ted Joanen (1982, 1983) were the first to report TSD in alligators. They performed experiments demonstrating that high temperatures (93°F, or > 34°C) produced all males, low temperatures (86°F, or ≤ 30°C) produced all females, and the intermediate temperatures produced a mixed sex ratio. Using switch experiments, they determined the thermosensitive period. In field experiments in Louisiana they discovered wet marsh nests to be cool and produce females, whereas dry levee nests were hot and produced males. Dry marsh nests were found to produce females at the periphery and bottom, and males were produced at the higher top center. The average sex ratio for dry marsh nests (and for nests overall in their study) favored females 5:1. All of the nests in their field study were composed of marshhay cordgrass.

At the time I was conducting another nest temperature study to ascertain the sex of the alligators in wild nests of various types of nesting materials and correlate the sex ratio to the incubation temperature for each nest, the study of TSD was in its infancy. TSD had been found to exist in some turtles, lizards, and a few species of crocodilians. It had been documented in Australia in the saltwater crocodile (*Crocodylus porosus*) and Australian freshwater crocodile (*C. johnstoni*). In these two crocodile species, bimodal TSD was known to exist, where females occurred at both low and high temperatures, and males occurred in between. Alligators were thought to be unimodal, with low-temperature females and high-temperature males. Then, when I and others began to get high-temperature females (one wild nest in my case with embryos having jaw and tail deformities (Hayes-Odum and Dixon 1997; Hayes-Odum et al. 1993); the others were laboratory studies with incubators, producing a couple of females per clutch at 91°F to 93°F [33°C to 34°C]), it was postulated that in alligators bimodal TSD did exist but at near-lethal temperatures.

Although I found that the type of nesting material could predict how warm a nest tended to be, there were exceptions. For example, common reed nests tended to be warmer than marshhay cordgrass nests. However, one giant reed nest at Murphree WMA was cool inside the nest and in the vicinity immediately around the nest. The coolness was believed to be the result of the nest being located extremely close to the water. Also, I found that cool, rainy weather could result in a substantial drop in the nest chamber temperature (fig. 3.34).

Wild nests can also contain hot and/or cool spots. Thus, zoos and crocodilian farming

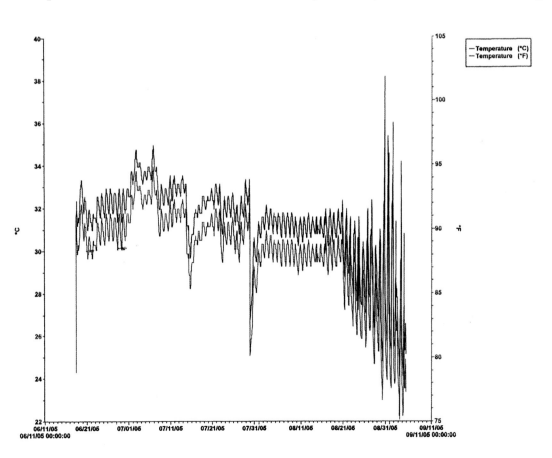

Figure 3.34. Cool, rainy weather at the end of July resulted in a substantial drop in the nest egg chamber temperature of this nest, which was being monitored.

operations can utilize incubators set at a specific temperature to maximize their chances of producing a particular sex ratio. Although wild nests moderate temperature changes within the nests, the temperatures still can fluctuate because of location, nesting materials, and climate change, resulting in a complex system that makes the sex ratio far harder to predict than for eggs incubated at a constant temperature.

However, sex ratio is not totally predictable, even with a constant incubation temperature. A phenomenon known as "clutch effects" seems to override TSD in determining the sex ratio of a particular clutch. One clutch of eggs incubated under the same conditions as another may have an entirely different outcome in its sex ratio. One cause of clutch effects might be attributed to yolk steroids, hormones that sometimes exist in the yolk as a result of the hormones circulating in the female alligator at the time of egg deposition. Yolk steroid levels can vary substantially from clutch to clutch, but within a clutch there is little variation. There is a substantial amount of research being done on yolk steroids, and the results have been conflicting about whether the steroids can cause clutch effects for TSD.

In addition to sex ratio, both temperature and clutch effects can affect growth and behavior. Temperature has even been shown to affect the number of stripes and amount of pigmentation. Hatchlings incubated at 91°F (33°C) have been found to have darker pigmentation and an increase in the number of stripes than those incubated at 86°F (30°C) (Deeming and Ferguson, 1989).

The hatchling sex ratio of alligators at a particular location may vary from year to year, and the hatchling sex ratio may be different from the sex ratio of adults in a population. Furthermore, since this is a long-lived animal, adults can vary substantially from each other in age. Aside from differences in hatchling sex ratios from year to year, the sex ratio of adults may be affected by differential mortality based on sex. Studies throughout the alligator's range have differing results concerning adult sex ratio, with a 50:50 sex ratio, a male bias, and a female bias all being reported.

ESD can influence the sex ratio of those animals with sex chromosomes, even human beings. There can be differences in the sex ratio at the time of conception (the proportion of each sex fertilized), number of live births (embryo survival by sex), and fitness/survival of offspring by sex due to the environment. In humans, there is a lot of research being done in this area. Males seem to be the more fragile of the two sexes as embryos and in the first few weeks of life. When conditions are unfavorable or there are fewer resources, females tend to be born. One hypothesis is that women who are healthy and live under good environmental conditions have a higher level of testosterone circulating in their bodies, which in turn supports the survival of male embryos.

Nest Predation

Fire ants, raccoons, and feral hogs are common predators of alligator nests in Texas. Fire ants are often found in alligator nests, but their presence does not mean that the developing alligators are doomed. The problem occurs when a rotten egg explodes, and an eggshell is damaged during the incubation period, or at the time of hatching the ant mound is located too close to the emerging hatchlings and a parent does not get them out in time. Newly hatched alligators stay at the water's edge and may be attacked by fire ants as well.

A nest predated by a raccoon typically has a single hole in its side, and there are eggshells scattered in the nest vicinity (Figure 3.35). A raccoon preying on an alligator nest is in danger of becoming prey to the female alligator that built the nest. This means that a raccoon may be aware of the presence of an alligator nest but not prey on it, either because it fears the alligator or its attempts are thwarted by the guardian female. However, I thought that a couple of nests predated by raccoons at Brazos Bend during my nest study on TSD may have been due to the human scent attracting the

raccoons to the nest site. Subsequently, rubber boots and gloves worn on visits to alligator nests were dipped in the lake mud and water to mask human odors.

Even if a raccoon is aware of the presence of an alligator at the den site, it may not be aware of the presence of a nest. Obviously, not all alligators are females, nesting is restricted to a few months out of the year, and a female may not nest for several years. Perhaps experienced raccoons have the knowledge that they should look for alligator nests at a specific time of the year, check locations where nests have been built in previous years, and be sure that the female is a distance away from the nest before they attempt predation.

The feral hog population increased at

Figure 3.35. An attempted nest predation, probably by a raccoon. Something other than the female alligator must have scared the predator away, as the nest had not been repaired. (Photo from the author's collection)

Brazos Bend State Park (and in many areas of Texas) in the 1990s. It became common to see areas near the foot trails that the hogs had rooted up. Nest predation by the hogs also started to occur. Hogs do not make tidy holes in a nest and extract the eggs like raccoons do. They destroy the entire nest and leave the nest area in a state of disarray. Also, since they often travel in family groups or herds, multiple hogs may be involved in predating an alligator nest.

Ted Joanen, Larry McNease, and Ruth Elsey (pers. comm.) noted that females added nesting material and reconstructed nests during the early incubation period. When the researchers opened nests and removed eggs, leaving the nest cavity open, the females repaired the nests. As the nesting period progressed, the females moved greater distances away from the nest. Two reasons for this affinity to the nest site seem apparent to me: nest integrity and predation risk. First, when the females are using green plant material for the nesting medium, it is somewhat unpredictable how the final nest structure will end up because the plants decompose. If the nest integrity is not totally predictable as the plant material in the nest degrades, then it is in the female's best interest for nest success to stick around and "patch" the nest as needed to protect the eggs from the elements. Possibly even more important is the predation risk. My observations of numerous red-eared slider nests being predated shortly after the eggs are laid suggests that perhaps the same is true of alligator nests, and I have noted this in several alligator nests.

Hatchling Alligators

Hatchling alligators are equipped with a temporary structure on the tip of the snout known as a caruncle, which allows them to pierce the shell membrane and eggshell (i.e., pip). The caruncle is not a true egg tooth, as found in snakes and lizards, because it does not contain calcium. It is made of keratin, the protein that composes structures such as hair, horn, scales, and claws. The shell membrane appears to be much harder to pierce than the eggshell, as loss of calcium from and microbial degradation of the eggshell assist in making the eggshell thinner and weaker than when it was originally laid. From my experience of watching a number of eggs hatch in captivity that spent much of their incubation time in wild nests, it appears that once the alligator's snout is able to emerge from the membrane, there is extra space freed up in the egg that allows the alligator to move its body enough so that the weakened shell develops longitudinal cracks and the hatching alligator is able to emerge rather quickly (fig. 3.36).

The yolk sac provides nutrition for the alligator during the course of the incubation period and for several days afterward. During incubation the yolk sac is separate from the embryo proper and attaches to the abdominal region via a stalk that has a rich blood supply. By the time of hatching the yolk sac should be drawn in to the abdomen of the young alligator and the abdomen fused shut. Sometimes this does not happen, and the yolk is still external at the time of hatching. If the hatchlings are truly full term, the vascularization in the stalk has diminished to the point where removing

Figure 3.36. (a–d) Once the egg membrane is pierced, the hatching process usually does not take long.

the yolk does not cause any problems. In these cases, the yolk may even become detached as the alligator emerges from the egg. However, if the alligator is not quite ready to hatch, it can bleed to death if the yolk and stalk are severed.

After hatching, the young alligators are found at the water's edge resting with their abdomens distended by the internalized yolk (fig. 3.37). Their bodies are soft and somewhat rubbery to the touch. They are at first black bodied with some white markings, the most noticeable being white dorsal crossbars. Within a few days, all of the white markings change to a yellow color.

After a year of age the body loses the rub-

bery characteristic and becomes more armor-like as the years pass and the bony plates in the dorsal surface develop. As an alligator ages, the stripes and other yellowish markings gradually decrease until it becomes an overall black color. However, the venter (underside of the body) remains light-colored, even in adults that are totally melanistic on the rest of the body.

After the yolk is absorbed, the young alligators are often seen in the water, swimming, feeding, basking on a log or other surface protruding from the water, or remaining buoyant at the water's surface. However, the young do spend a substantial amount of time on land. Hatchling alligators are often more visible in the water than on land, even with camouflage provided by logs, duckweed, and other plant material. The height of terrestrial vegetation, coupled with the loglike appearance of the youngsters, makes them blend in quite well. Feeding may take place during the day but is more common at night. Much of the day is spent basking, and the youngsters in the pod tend to be in close proximity at this time. They may even pile up on top of each other to bask (see fig. 3.4). During foraging activities at night, the hatchlings tend to separate themselves at the greatest distances.

Juvenile Sex Ratios and Growth

Although it is easy to ascertain the sex of an adult alligator by probing the cloaca to check for the presence of a penis or the much smaller clitoris, it is very difficult to determine the sex of very young alligators except by sacrificing them and removing, sectioning, and staining the gonad for microscopic examination to see if a testis or ovary and oviduct are present. However, young crocodiles are presumably relatively easy to sex via probing the cloaca.

A few larger juveniles from my mark-recapture study were able to be sexed via finger probing of the cloaca, but I was unable to get enough data to substantiate any trends in sex ratio. Laparoscopy, in which a small surgical incision is made into the abdominal cavity, was attempted. However, no macroscopic gonadal differences were seen. A testosterone immuno-assay technique was tried on a limited number

of samples taken from the Houston Zoo and Murphree WMA and appeared to exhibit variation in results that could be due to sex-related differences.

Perry Viosca (1939) detailed a method he developed using the size, shape, and number of scale rows in young crocodilians to ascertain the sex. He claimed that males had three definite rows with a rudimentary fourth row on either side of the cloaca with scales that tended to be larger, subtriangular, and platelike. Females had four definite rows with a rudimentary fifth row and smaller, more rounded, beadlike scales. I tried to use this method on both adults and juveniles. I even photographed a substantial number of cloacas of adult alligators taken on the first hunt in 1984 to see if the method had any merit. It seemed to work up to a point (50 % –80 %), but at other times it just was not valid. I now think that this method involving the size and shape of scales and number of scale rows really may be related to the rate of development, which is related to the incubation temperature (as is the degree of pigmentation and the number of stripes), which is in turn related to the sex.

Ruth Elsey at Rockefeller Refuge in Louisiana showed me a method to ascertain the sex of alligators as young as 18 months with a nasal speculum to view the phallus (penis or clitoris). It seemed to work well in those animals where there was sufficient differentiation of the phallus. I did sex what I thought was an obvious male this way. Years later, it turned out to be a high-temperature female, as evidenced by it laying eggs. This instance emphasizes that temperature, rather than sex, may be responsi-

Figure 3.37. A pod just after hatching.

ble for characteristics that could be statistically linked to sex if the individuals of one sex were linked to only one temperature regime. Specifically, high-temperature females rather than low-temperature females may be the most likely to show extremes in differences that manifest themselves in physical attributes.

In the first four years of life, the growth rate is quite rapid. As previously mentioned, it averages out to about a foot per year. From my mark-recapture study, I found that the greatest period of growth in juveniles was from July to August, when they increased in total length by 18%. From May to August, juveniles increased in total length by approximately 50%.

Parental Care

Generally, we consider parental care in an egg-laying animal to constitute any care given to the eggs or resulting offspring after the eggs are laid. In other words, if an animal simply lays its eggs and leaves, then it does not provide any parental care after that point. The choice of nest site, how a nest is constructed, and so on may be very important as to the successful hatching of young, but this is not parental care in the sense discussed here. If the animal guards its eggs, assists in the hatching, and is involved in guarding and/or caring for the offspring after hatching, then any of these activities singly or in combination constitute parental care. In lower vertebrates, parental care is well developed in a number of species. For example, cichlid fish guard eggs and newly hatched offspring. They also carry newly hatched offspring in their mouths back to the "nest site" if they stray into an open area where they could be eaten by other fish. Amphibians have a number of known instances of parental care as well, such as the terrestrial Rio Grande chirping frog's (*Syrrhophus [Eleutherodactylus] cystignathoides campi*) territorial guarding of its eggs. Unusual cases occur, such as the male Suriname toad (*Pipa pipa*), after fertilizing the female, takes the eggs and presses them into her back, where they hatch out.

In reptiles as a group, parental care often appears less advanced than what is seen in lower vertebrates or is absent altogether. Somma (2003) has done a literature survey of parental behavior in reptiles aside from the crocodilians. His work provides extensive literature citations and breaks the behavior into discrete categories. Perhaps the most well-known noncrocodilian parental care in reptiles is the "brooding" behavior of pythons, in which a female coils around the eggs, shivers, and is able to raise the temperature of the eggs by several degrees. Some monitor lizard species lay their eggs in termite mounds and return at the time of hatching to liberate the young from the mounds, which were sealed off by the termites. In at least two species of another type of lizard, the skink (*Eumeces* spp.), the mother is known to lick the amniotic fluid off the bodies of its young after they hatch. Fewer than 5% of noncrocodilian reptile species are known to exhibit parental care. Many reptiles are relatively small, secretive species and live most of their lives involved in behaviors unseen by the human eye. Consequently, the surface has barely been scratched on the subject of parental care by reptiles, an area that needs more research. Also, the size disparity between human beings and many reptiles may in fact be one reason for behaviors such as defense of the nest or young to be underreported. Is it plausible that a small reptile is going to react in the same manner to a gargantuan human as it would to a much smaller typical predator of its eggs or young?

Crocodilians exhibit highly developed parental care comparable to that of birds or to what we infer about parental care in the extinct dinosaurs. It is one of the main attributes used to separate crocodilians and dinosaurs from the other reptiles and couple them with birds. Given that highly developed parental care has been noted in vertebrates other than the crocodilians, birds, and mammals (i.e., the "lower" vertebrates), is parental care more an adaptation for species survival rather than a characteristic associated only with the highest, most advanced vertebrate groups? Because it is one of my major research interests, I am not trying to downplay the amazing phenomenon of parental care in crocodilians. Rather, I am trying to put the subject of parental care in a balanced perspective. Furthermore, although crocodilians are very large, relatively conspicuous animals, especially in relation to their social and reproductive behaviors, certain aspects of their parenting were not accepted as factual until the 1970s. Others are still not.

Nest Defense

Crocodilians are known to guard their nests from potential predators. Behavior at a nest when humans approach can be extremely variable. It may be that the female is never seen; the female is in the vicinity but turns away from the nest and splashes in the water; the female lies in the water beside her nest with just her head showing and makes no aggressive moves; or the female is very aggressive, goes toward the nest, climbs on top of the nest, and lunges off it with her mouth open (fig. 3.38). In the most extreme circumstances, the female chases the intruders a considerable distance.

A study by James and Marilyn Kushlan (1980) demonstrated that there was a succession of behaviors in nest guarding, and a particular animal might stop at some stage of the sequence and withdraw. They named the individual behaviors and chronicled the behavioral sequences of the nest defense. They also showed, through the use of a stuffed raccoon

Figure 3.38. Aggressive females commonly respond to nest intruders by going toward the nest, climbing on top of it, and lunging off it with the mouth open.

and a dummy resembling a human, that part of the behaviors constitute bluff and others are "serious business." The stuffed raccoon (in life both a natural predator of alligator eggs and an adult alligator prey item) was grabbed from the line with a hard bite and eaten after an initial open-mouthed approach with hissing. The much larger human dummy was given a "mock bite," in which the alligator bit down lightly and released, rather than an immediate hard bite when contact was made. Only if the human model did not retreat was the "hard bite" resorted to, after which the alligator withdrew.

When I conducted a nest study at the Murphree WMA during 1987 and 1988, one female out of eight was aggressive toward me when I approached her nest. The airboat did not even seem to phase her. We would try to physically position the airboat between her nest and her so that I could access the nest. Her nest was visited twice (once to insert a nest temperature monitor and then to retrieve the monitor and remove the eggs), and the same behavioral sequence occurred on both occasions. However, at Brazos Bend State Park during the same period, no females attempted to make aggressive moves toward me. They either would position themselves in the water near the nest with just their heads visible, or I would hear a splash in the water in the vicinity and turn around to see part of the alligator's body come out of the water for a brief instant. The splash in the water seemed to be an attempt to divert attention away from the nest, much like the "broken wing" behavior used by birds.

After the study was completed at Brazos Bend involving the collection of eggs from nests, I continued to collect data there on nest attributes and the size and behavior of attending females for a number of years. I noticed a change in behavior in the alligators over the next few years, in which an increasing number of females became aggressive to humans in the vicinity of their nests. Females that formerly would lie in the water next to the nest while it was being examined began to hiss and lunge at the human invaders. It finally reached the point that all nests in the high-traffic areas of the park were actively guarded by aggressive females. Only the remote areas still had nests in which the female was never seen or was seen to splash in the water in the nest vicinity but not go near the nest. Ruth Elsey related to me that she had also noticed an increase in aggressiveness of nesting females at the Rockefeller Refuge that had become used to humans. It appears that alligators that become used to humans lose their fear enough to aggressively guard their nests against them. These animals are not the same as habituated alligators (see chapter 6), since the aggressive behavior is restricted to nest guarding. Also, the nest guarding does not occur until the eggs are actually laid, so the presence of a guardian female at a nest is indicative of eggs being present. It is also noteworthy that a number of literature accounts seem to indicate that alligators that are harassed by humans do not actively guard their nests.

Nest Maintenance

After the eggs are deposited and the female completes the nest, it is not uncommon for her to "tidy" it up on a regular basis. She may smooth or shape it, bring in more nest material, or place a branch or other object on the nest exterior. I have found unusual objects on the nest such as beer cans, plastic bottles, a camera lens cap, skulls, antlers, and shells. Are these ornaments, specifically collected to place on the nest, or merely objects present in the nest site area? The alligator may also void urates, the semisolid waste material from the kidneys composed of uric acid with some liquid, on the nest surface (fig. 3.39).

It has been alleged that by urinating on the nest, the female is keeping it moist. The watery, semisolid waste material can have a significant amount of water in it and could make the nest moist in a limited area in the superficial lay-ers, but evaporation could take a big toll on the moisture that remains for any period of time. The outside layers of the nest tend to dry out in the absence of rain, with only the nest cavity containing the eggs being moist. The alligator puts wet nesting material into the nest, and when she seals it up, it is insulated enough to stay moist. However, alligators do lie on top of their nests from time to time after they have been in the water. It is sometimes possible to see a wet outline of the alligator's body on a nest after she has recently crawled on top of it. It has been suggested that this temporary wetness may exert a cooling effect on the nest. It has also been postulated that urates help ward off predators by indicating the recent presence

Figure 3.39.
Urates secreted
by the female
on top of the
nest may serve
to discourage
potential nest
predators. (Photo
from the author's
collection)

of the female, and this is quite possible, especially when they are fresh.

Nest Opening

Although Joanen (1969), McIlhenny (1935), and Reese (1915) all mentioned alligators liberating the young from the nest, none actually stated that they had personally seen a female open a nest. There were reports of nest opening in other species also: the Nile crocodile (*Crocodylus niloticus*; Cott 1961; Modha 1967), and the spectacled caiman (*Caiman crocodilus*; Alvarez del Toro 1969). However, Neill (1971) refused to acknowledge these accounts and said he observed that nests of females subjected to heavy hunting pressures still hatched without presumed help, and although he did believe that females guarded nests, he did not think that nest opening took place. Ogden (1971) wrote of nests being opened by American crocodiles (*C. acutus*), but not until he published a paper with infrared photographic proof by Singletary, as well as a written description, of an American crocodile opening its nest, carrying eggs and young to the water, and assisting in the hatching of the unhatched eggs did it become "fact" that crocodilians did open nests to liberate their offspring (Ogden and Singletary 1973; see also Hunt 1987).

I have noticed, as the time gets closer to the end of the incubation period for alligator nests, that it is not uncommon to see an imprint in the shape of the female's head on the nest surface. Literature accounts mention frequent visits by the female crocodilians to the nest near the end of incubation and the female resting her head on the nest as though she is trying to hear the offspring or feel their movements. Lee (1968) postulated that the young communicate among themselves, as well as to the female, by their movements inside the egg about two weeks before hatching and then by vocalizations just prior to hatching. He thought that this communication served to synchronize hatching and that the movements might even aid in regulating embryo growth.

The young within the eggs grunt softly at first, then change their vocalization to a loud chirp, which can be heard several yards/meters away from outside the nest. This should signal to the female that it is time to open the nest. I know of two nests at Brazos Bend where chirping was heard when humans walked past them. In the first case the female did show up to open the nest. In the other instance, it was in a remote area of the park and the female was not seen. We opened up the nest and found that the young had started to hatch inside the nest and fire ants were present throughout the nest cavity. Of the 39 eggs, 14 were rotten or contained dead embryos. The rest we assisted in hatching and dipped them in the water to wash off any ants. They were released in the shallow water. It is unlikely that they would have gotten out of the nest without our help and would have been killed by fire ants before the female returned.

The parents can also open the nest prematurely before at least most of the offspring are ready to hatch, as an instance at Brazos Bend demonstrated. A nest was opened after about 75 days of incubation by one or both parents before most of the offspring were ready to hatch. We had been monitoring this nest closely, as it was conveniently located, and we hoped to film a nest opening. The female had been visiting the nest frequently and an outline of her head had been clearly visible on the nest surface for several days. We believe that she was listening for signs that the young were ready to hatch. We had last checked the nest late the previous night and had rechecked it early in the morning to find that it had been opened. Five empty eggshells and four dead embryos, which were not quite ready to hatch and possessed a large amount of unabsorbed yolk, were conspicuous at the nest site. Three eggs were cracked and were being entered by fire ants. As a result they had to be opened but were not quite old enough to survive.

Twenty-nine intact eggs, removed from the nest and incubated artificially, hatched 4 to 12 days later, and the young were released at the nest site. However, there was no sign of the hatchling belonging to that fifth empty eggshell. Could one or more offspring have been ready to hatch and taken to the water by the parents? The female and another larger alligator thought to be a male were swimming around in the water in a specific area away from the nest site. They were very alert and seemed to be patrolling the perimeter of this area as though they were protecting something, possibly one or more hatchlings. This nest was in heavy shade, and perhaps there was a hot spot in the nest where at least one egg was incubating at a faster rate than the others. The cracked eggs were out of the nest cavity, lying on top of the nest, and were presumed to have been cracked by one of the parents starting to open them before realizing that they were not ready to hatch.

Egg Opening and Egg/Offspring Transport

Most crocodilian species are now known to not only open the nest but also to assist in opening eggs and transporting newly hatched individuals to the water's edge. The parent opening the nest gently rolls each unhatched egg in the roof of the mouth with its tongue to crack the eggshell in order to liberate the young alligator. The hatchlings may be carried to the water in the parent's mouth. The parent may also "wash off" the eggshell pieces and amniotic fluid from the hatchlings by taking each one in its mouth. Jim Tamarack, who worked for the New York Zoological Society at its St. Catherine's Island, Georgia, sanctuary for endangered species, played alligator parent to a nest that he relocated to a pond he built for the purpose of studying juvenile alligator behavior. He carried the babies to the water in his mouth after they hatched and noted the reason that the young alligators do not get injured by the parents' sharp teeth is that they go totally limp when they are picked up.

Defense of Young

The mother stays with the young for a period of time ranging from months to years (fig. 3.40). Young alligators often rest on the head or back of a parent both on land and in water (fig. 3.41). They may use the adult's back as a basking area, a diving platform, or even a boat of sorts to move across the water (fig. 3.42). When a juvenile gives the high-pitched "umph umph" distress call, the female comes to the aid of its offspring. She may lunge at an intruder and/or pick the youngster up in her mouth to protect it.

Figure 3.40. The mother stays with the young for a period of time ranging from months to years.

Feeding of Young

An alligator feeding has been photographed, yet most herpetologists either are unaware of its existence or believe it to be cases of tolerance by a parent in having youngsters pick food out of its jaws. The most detailed historical account of which I am aware is from McIlhenny:

> I have seen mother alligators catch large fish, large snakes, and turtles in their jaws and crush them to a pulp, holding them at the surface of the water between their jaws, so that the young could gather bits of food from the crushed flesh. The young would grasp the food in their mouths and with vigorous shakes of the head tear off bits of it which they would swallow exactly as the old ones do, by raising their head in the air and gulping down the food. I find in going through my notes on the alligator that I have witnessed this method of young alligators getting food on eight different occasions—four times on garfish, twice on snakes, and twice on turtles. (1935, 54)

Biologist Larry McNease related a detailed observation to me that he witnessed at the Rockefeller Refuge in Louisiana involving a

a

c

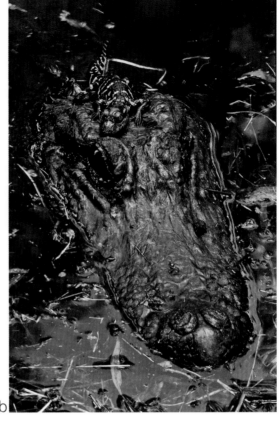

b

Figure 3.41. (a–c) Young alligators often rest on the head or back of a parent.

captive male alligator that adopted young alligators whose mother had died. The penned alligators at the refuge were fed skinned, ground nutria that was partially frozen. The male crushed the semifrozen meat in its jaws and allowed the babies to eat the meat from his mouth.

A photograph was taken in 2001 of a female Siamese crocodile (*Crocodylus siamensis*) at St. Augustine Alligator Farm with a skinned nutria in her mouth, which was being fed upon by her young. This was the May 2001 crocodilian.com's Pic of the Month. The text that accompanied it said that she sat for over an hour with her mouth slightly open, letting the young feed on the meat. It also noted that this behavior had been reported in captive broadsnouted caimans (*Caiman latirostris*) by an experienced keeper and in wild Orinoco crocodiles (*Crocodylus intermedius*) by a researcher. The possibility was brought up that this could be an opportunistic case of scavenging by the young crocodilians rather than a deliberate behavior by an adult to feed the young.

Figure 3.42. The adult's back may serve as a basking area, a diving platform, or even a boat of sorts to move across the water.

Indirect evidence of either parental feeding or scavenging by young crocodilians showed up in my food habits study. Of the seven mammals I found in juvenile stomachs, four were in alligators less than 2 feet (0.6 m) long. One of these alligators had hair from an adult rabbit in its stomach. Obviously, this youngster most likely was incapable of preying on an adult rabbit.

Why might an adult feed its young mammalian prey, when it is known that mammals are not a common dietary item of juveniles? One possibility involves getting the young

crocodilians used to the smell/taste/feel of the mammals so that they will have a tendency to prey upon them in the wild to make a leap from invertebrates (chiefly insects) to mammalian prey. Perhaps we should look at this as a means to bridge the crossover from largely invertebrate to vertebrate prey. In McIlhenny's observations, he never noted feeding of the young by a parent involving a mammal. They were fed gar, snakes, and turtles. We seem to be looking at the parent introducing vertebrates to the offspring before they are ready to "take them on" as prey. The scenario with the Siamese crocs at St. Augustine Alligator Farm occurred when they were past the young hatchling stage, and this was also the case with the juvenile alligators whose stomachs I pumped.

Role of Males in Parental Care

It is interesting that although a female alligator may have offspring from the same clutch of eggs sired by more than one male, males are known to participate in parental care activities. The factors that determine if, when, and in what manner they participate are not known. From my observations and the literature, nest guarding seems to be the job of the female. The male crocodilian sometimes shows up for the nest opening and at times will open the nest instead of the female. He may also crack open the eggs to free the young and carry them to the water.

The instance at Rockefeller Refuge of a male alligator adopting young alligators after their mother died is not the first on record. McIlhenny (1935) found 28 young alligators in a pen with an 11.5-foot (3.5 m) male after the death of their mother. They basked on top of his head and back and stayed with him in his den for the winter. Philippe's photo of a male and female alligator with the offspring in the grass in front of their heads hints that male involvement with young may not be uncommon after hatching (fig. 3.43).

Figure 3.43. A male and female alligator with the offspring in the grass in front of their heads. Male crocodilians are known to be involved in nest opening and cracking open eggs to liberate the young. It is not known to what extent males are involved with the offspring after hatching.

Crocodilian Parents Feeding Offspring

Two wild ideas that need further investigation:

Could parental feeding of young also be a response to limited food resources for the young at a nursery area? I can think of an instance at Brazos Bend where an alligator's nest site was not an appropriate location for the nursery because there was little food available for the young. It was an area with an intermittent stream, and it was unlikely that there was enough invertebrate food to keep the young fed. The female probably had to go elsewhere to obtain most of her food. This does not solve the food problem for her offspring, unless she brings back meals for her young. The next time such a nest appears at Brazos Bend and hatches successfully, it would be interesting to observe interactions between the female and her young to see how the young obtain food.

Two instances of pet crocodilians bonding with the family cat provide some interesting possibilities regarding parental behavior toward cats, especially feeding of the young. From these accounts, there appear to be some behaviors that parallel female crocodilians providing care for their young. In one case a pet cat and caiman slept together on a regular basis, with the cat resting its head on the caiman's back. In the summer months, the cat and caiman roamed the backyard together, where the caiman had a plastic swimming pool. The cat and caiman also ate together in the kitchen, and the cat many times ate pieces of fish from the caiman's mouth (Anonymous 1962). The second case involved a pet alligator that bonded with the family cat and allowed it to lick food off its face (Sullivan 1994). Could certain aspects of the cats' overall behavior elicit some sort of parental care response from the crocodilians?

John Bradshaw (2013) has looked at cat behavior and suggests that the domestic cat does not have the same awareness as dogs do that other species are a different species from themselves. As a biologist and a cohabitant of cats since childhood, I find this concept hard to accept and offer a paraphrase of the famous line from the movie *Gone with the Wind* to suggest what cats really think: "Frankly my dear, I don't care what species you are." My thoughts are that cats are aware of different species, but it is not a big deal to them. Check out some YouTube videos showing that cats do not care about size or species when chasing much larger predators away. Some videos show pet cats successfully chasing away wild alligators and bears.

Communication

Crocodilians communicate with each other and with other species (including humans) through sounds (both vocal and nonvocal), visual elements (movements, postures, and subaudible vibrations), and odors. Specific behaviors, such as guarding or patrolling a territory, attacking an opponent, bellowing, mating, nest guarding, and other forms of parental care have been described earlier. Here I look at the specific elements of behaviors and what each means in the context in which it is used. I have observed all of the behaviors documented and use terms from studies by Garrick and Lang (1977), Garrick and Herzog (1978), and Vliet (1989).

Vocalization

Alligators begin communicating using vocalizations before they even hatch. They may chirp softly near the end of incubation and then more loudly at hatching time so that it can be heard some distance outside the nest to signal a parent to open it. The chirp is an "umph" sound if given in a single note and "umph umph" if expressed as a double note. Juveniles chirp to each other, presumably to bring or keep the pod of youngsters together (i.e., group cohesion). The juvenile voice is high pitched and becomes an even higher-pitched and louder "umph umph" if the juvenile uses it as a distress call and is trying to get an adult to come to its aid. Older juveniles start to use a hissing sound if they feel threatened or cornered, and this holds true for adults in general. However, I am aware of an instance when a large subadult/small adult used a distress call as it was being cannibalized by a larger alligator.

Females are known to communicate with their young with grunts, especially to round them up. It is possible that they also signal to the young about potential dangers such as predators, or if food is nearby, such as an aquatic insect swarm or school of fish. The young tend to disperse in the evening and gather in the day to bask. They feed at night or when opportunistic situations (such as aquatic insect swarms) present themselves during the day. In active nest guarding, hissing is an important component of the female's behavioral sequence. The bellow-growl is a lower-intensity single bellow made by females that appears to be associated with male behaviors, such as head slaps or attempts at courtship. This seems to be an announcement that a female is present and wants to be left alone.

"Chumphs," coughlike sounds associated with both sexes during mating, are produced by air being expelled through the nostrils. Both males and females give low growls associated with aggression.

The bellow is the most impressive, noteworthy vocalization associated with alligators and, as previously mentioned, is performed by both sexes and can carry quite a distance. Each individual's voice is unique. One alligator bellowing can stimulate others to bellow. Bellowing occurs most of the year, except during winter. However, it is most frequent during the breeding season and seems to be associated with mating and territoriality. At Brazos Bend, bellowing has been known to start as early as the end of February. But mid-March through May is the most common time for bellowing in Texas. At times, there has been a smell associated with the bellowing at Brazos Bend that is like rancid icing. Since the chin glands are known to be sometimes everted during bellowing, it seems likely that this odor emanates from these glands.

The largest bulls (largest males) demonstrate the most spectacular bellowing displays. Due to their large size and ability to displace a much larger amount of water, the large bulls can come forward, up, and out of the water to a greater extent than the smaller alligators. Their large head and upper body are prominent out of the water and visually make a great impact on any human onlookers and certainly give a large dominance display to other alligators (fig. 3.44).

Although both sexes bellow, there is one component to the bellowing repertoire that is "males only." This component results in the water "dancing" around the alligator's body due to subaudible vibrations being produced. In this sequence, the alligator raises its head, gulps air, and drops lower into the water (fig. 3.45). Then the alligator gets into a position that has been termed the HOTA (head oblique tail arched), in which the animal raises its head to an angle of 30° or 40° with the tail arched. Then the infrasonic signal known as the "subaudible vibrations" is produced, and the water "dances" (fig. 3.46). Immediately thereafter, a throaty, roaring bellow occurs.

Nonvocal Sounds or Acoustics

The head slap (sometimes termed "jaw clap" in the literature) takes place at the surface of the water/water-air interface. The head is above the water with the lower jaw barely visible. The result is that the jaws close so that the biting motion is at the surface of the water, with a popping sound followed by a splash. This is a sound heard both in the air and underwater. Other alligators may approach or retreat after this display. This is a dominance display used by both sexes in the context of dominance and/or territoriality.

Subaudible vibrations, the signals made by male alligators prior to bellowing (when the

Figure 3.44. Large male alligators are quite impressive when they bellow.

Figure 3.45. An alligator has gulped air in preparation to emit a bellow.

water dances), are very low-frequency sounds. At close range these vibrations resemble distant thunder, and it is thought that they could travel quite a distance beneath the water. Narial geysering is usually associated with courtship and mating and is the "blowing bubbles" noise.

Visual Signals

The snout lift is a signal that signifies subordinance, submission, and/or nonaggression. It is associated with mating behavior to demonstrate the willingness to mate rather than aggression. It is also associated with subordinate (especially subadult) alligators passing through the territory of a larger/dominant alligator.

The tail wag (not associated with feeding behavior) is correlated with defensive postures to indicate aggression. The HOTA posture is usually associated with bellowing. Inflating the body (inflated posture) is an aggressive posture, often associated with nest guarding, bellowing, head slapping, territoriality, or being cornered. In addition to inflating the body (which makes

Figure 3.46. The water "dances" as this male alligator bellows.

it look bigger and more formidable), the head and all, or a substantial portion of, the back are displayed.

Odors

Paul Weldon has been a pioneer in the field of glandular secretions and skin components in reptiles (Weldon, Flachsbarth, and Schulz 2008). He was the first person (at least in the United States) to begin studies to examine all types and species of reptiles to find out what composed their gland and skin secretions. Obviously, the technology had to be developed in order to conduct these studies, but he was the first person to take advantage of the technology. He asked zookeepers around the coun-

try to accommodate him in his research efforts. Some of his studies involved having zookeepers collect snake skins and send them to him. Other studies involved his actively "milking" secretions from reptiles. Zoo personnel also conduct research and try to accommodate others doing worthwhile research. These more "active" studies meant that they had to restrain the animals while Weldon was obtaining the secretions. I am sure that at times it was somewhat difficult trying to restrain a reptile that did not want to be "milked."

Crocodilians have a pair of glands in the chin/throat region (gular glands) and in the cloaca (paracloacal glands). Weldon was at Texas A&M University while I was doing my graduate work there. He gathered gland secretions from some of my juvenile alligators. I remember the sickeningly sweet smell of something like Hi-C orange drink filling my nostrils as I was holding one of the alligators while Weldon was taking the samples. He then obtained glandular secretions from adults and from a number of other crocodilian species. He believes that the gular glands are used to mark territory and suggests that female alligators use their gular glands to scent-mark nest sites by rubbing the ground with their lower jaw. The paracloacal glands are believed to produce pheromones used in mating and nesting activities. He has suggested that the secretions indicate aggression by adults, and he observed juvenile alligators in the wild fleeing an area after thawed paracloacal gland secretions from adult males were poured into the water. His analyses suggest that there is variation in these glandular secretions based on size and sex. I think that my observations of a rancid icing smell (a sweet rancid oil smell) made by an adult during bellowing and the Hi-C orange drink smell were made by the same type of glands and that the age of the alligators made a difference in the odors.

Some Closing Thoughts on the Natural History of Alligators

Alligators have an interesting and complex life history. When alligators and other crocodilian species were extremely scarce, relatively little research was being done on them by a handful of scientists. After several species rebounded and rarer species increased in number, studies became more numerous, and the number of scientists studying crocodilians increased to the hundreds. Historical accounts of crocodilian behavior have been scrutinized and been proven, disproven, or explained in greater detail. Advances in technology have also contributed to an increase in the types of research being conducted on crocodilians. Although a lot of information has been published on alligators and their relatives, there is still a lot more to be discovered and documented on these fascinating creatures. I hope that some of the discussions in this chapter and the questions I have brought up will inspire researchers to contribute further knowledge to what we know about the alligator's life history.

4 Alligator Tales

Legends and myths, frequently based on partial truths or some kernel of fact, often changing and growing by word-of-mouth transmission over time (depending on the imaginative powers of the raconteur), do reach an enormous number of people. Often subtle and recognized, the pervasive influence of myth and legend has subconsciously affected the very ways in which we think about alligators. It is good to be aware of such factors as we form opinions or draw conclusions about this primordial southern saurian.

—Vaughn L. Glasgow, *A Social History of the American Alligator*

From time to time, I get e-mails with spectacular photographs of larger-than-life alligators, in terms of size or feats that they are caught in the action of performing. These types of e-mails get circulated to the general public as well, and I am asked about their validity. I also hear tall tales about alligators that I know better than to believe, but I realize that some people are gullible enough to swallow hook, line, and sinker. I thought it would be useful to go through some Internet myths to set the record straight and to help educate the public about what to believe and what to discard as absurd or unlikely, no matter what the source.

I also have heard some interesting stories that I would like to share involving Texas alligators. The stories of alligators in places where they are not supposed to be not only give range or habitat extensions of alligators but can be potentially helpful when an alligator shows up where it is not expected to be.

Giant Alligators on the Internet

There are a number of alligator photos, most notably of extremely huge alligators, that get circulated around the Internet. A good source to check on these for accuracy is Rumor Has It and www. snopes.com. This website allows you to search and submit photos, videos, and written proof to prove or disprove so-called urban legends. The most common ways to misrepresent a photo concern the actual size of the alligator or the geographical location. Also, a photograph can be doctored to display something that never happened. For example, there is a photo of an alligator farm pond (not in Texas) showing a parachutist getting ready to land into water with alligators waiting onshore. The original photograph from the alligator farm was submitted to Snopes without the parachutist to show that it was a fake. Presumably the prankster got the idea from a Gary Larson cartoon.

The Construction-Site Alligator

This has to be the most widely circulated alligator hoax on the Internet and has appeared intermittently for a number of years. It typically is said to represent a very specific area in the South where alligators are found (and is usually e-mailed to people in at least that state). By specific, I mean down to the town and road. I am assuming that the perpetrators realize that the locality is fake or just think that it would be fun to "change the locale" to their neck of the woods and then send it on.

The photograph shows two large culvert drainage pipes that fill up the background. The alligator in the foreground is tied up with rope, and its mouth is taped with duct tape. Two construction workers are shown standing in the mouth of one of the pipes, and the other three are between the pipes and the alligator. The alligator looks bigger than it really is, since it is in the foreground, and the pipes are huge. The effect makes the construction workers look very small and the alligator very large. It was reputed to be 18.5 feet (5.6 m) in length and weigh in at nearly 2,100 pounds (952.5 kg).

Crocodilian biologist F. Wayne King noted that the photo was first circulated on the Internet in June 2002 with a caption that it was killed at the construction site for the Florida Convention Center (which it was not). The alligator was killed by Billy Harter in Hillsborough County, Florida, and skinned by Mike Fagan. It was 13 feet, 5 inches (just over 4 m), but no weight was recorded. This information was reported to the Florida Fish & Wildlife Conservation Commission, since any alligators killed in the state must be reported to that regulatory agency. Also, all alligators killed in the United States are under CITES (Convention on International Trade in Endangered Species of Wild Fauna and Flora) regulations and must have a CITES tag, which has paperwork associated with it.

In 2009–2010 in Texas, a twist occurred to this spoof with a second photo being sent in conjunction with it. The second photo was of rattlesnakes (supposedly 47) in a drainage pipe. Both the alligator and snakes were reputed to have been found by a Texas Power & Light crew in various localities. The latest that I received (January 7, 2010) said that the photos were from Hallettsville Airport. It turns out that Snopes.com first got these photos as "a package deal" in August 2002, supposedly from the Florida Power & Light crew. According to Rumor Has It, the rattlesnakes turned out to be desert rattlesnakes, identified from the California Bureau of Land Management (letter included), which had access to the original photo.

The Giant Alligator Suspended from a Backhoe

This alligator really was killed in the backyard of a home at Bar X Ranch near West Columbia, Texas, in April 2005. The photograph, taken by the *Facts* newspaper photographer Val Horvath, was accompanied by an article stating that the alligator was 13 feet, 1 inch (approximately 4 m) long and was killed by game warden Joe Goff. The homeowners, Anita and Charlie Rogers, were interviewed in the story. The photograph made its way to Florida in 2006, and the alligator was listed as being 23 feet, 1 inch (approximately 7 m) long. The names of the homeowners and the game warden remained the same, but the locale was changed to a backyard in Florida. The photograph makes the already-large alligator look larger, since the alligator is in the foreground with its head turned toward the camera and the game warden is in the background appearing smaller than he really is. This photograph and story are listed on Snopes.com.

The Toledo Bend Alligator

This photo and caption allege that a 13-foot (3.9 m) alligator was killed in Toledo Bend, Texas. This message was sent as an e-mail to Bob Brown, chair of the Department of Wildlife and Fisheries Sciences at Texas A&M University, in October 2004. K. J. Lodrigue, a Texas A&M University graduate student who had been a TPWD biologist, disputed the locale given. He said that an alligator of that size in that area would have been illegal to kill. The photo shows an alligator strung up and with a man beside it (supposedly the hunter) with

a crossbow used to kill it. The alligator was in the foreground, which made it look larger. It was said to be 12 feet, 9 inches (3.6 m) in length. This is a reasonable size, but the locale is questionable.

Dead/Not So Dead Alligators

Dennis Jones of Texas Parks and Wildlife shared two stories of alligators believed to be dead that were not. One is a personal story. These are told in Dennis's words.

A Game Warden Responds to a DOR (Dead on Road) Alligator That Wasn't Really Dead

Remember the game warden who responded to an alligator call regarding a roadkill in the late '70s/early '80s? Alligators were then strictly regulated, and when he arrived at the scene, there was a crowd of people around the 10-foot alligator. Kids were sitting on top of the alligator getting their pictures taken. As the game warden moved the alligator to get a rope around it to move it off the road, the alligator turned and bit him on the knee. Pretty severe damage occurred to the game warden and Dr. Red Duke, the ER surgeon at Ben Taub Hospital, saved his leg by not closing the wounds immediately and debriding the wound regularly until the danger of infection had passed. I don't remember the warden's name, and this story is how I recall it being told.

The Brazos Bend State Park "Not So Dead" Alligator

In February of the mid- to late '90s, the park volunteers came to me and told me of a dead alligator at Creekfield Lake. I asked them a few questions about the alligator and told them, "If the alligator is not stinking and having flies buzzing around it—it is not dead." They continued to insist that I come and see the alligator. I went to the lake and saw the alligator about 40 feet off the water's edge. The alligator's head was under the water as well as its tail. Its midsection was out of the water, likely resting on floating vegetation. Someone had very accurately tossed a large, round stone onto the alligator to test its live or dead condition, and the stone was still resting on its midsection. I watched him for a long time to see if he would take a breath but saw no evidence of respiration. Furthermore, the alligator's right forelimb was sticking up into the air at a rigor mortis–like angle. I thought to myself and out loud, the volunteers were right: this was a dead alligator.

At that point, I thought I would like to collect the alligator and do a cursory necropsy for the growing audience and later recover the skeletal material for reassembly. I donned my chest waders and walked into the lake. I had to push apart the heavy vegetation and was ever concerned that the water was getting too deep for my waders. So far I had been fortunate, but the next step could bring the cold February water rushing into my waders, and that would be very undesirable. As I pushed forward, I noticed the floating vegetation was closing in behind me, meaning that I would have to part it again on the way out, pulling an 11-foot alligator. Why and how did I get into this situation? Stupidity or "scientific curiosity" may have been my motivation, or a combination of both.

When I got about five feet from the alligator, I could see how large he was, at least 11 feet, maybe more. I decided I would conduct one more test to determine if he was really dead. I broke a small, three-foot branch off a tree nearby and gave the dead alligator a gentle push. Nothing happened. I pushed again just to be sure, and something strange came to me. The skin of the alligator pushed in too easily. . . . If the alligator was dead and bloated (which I suspected was the cause of his midsection floating), his skin would be taut and not so giving. As I was processing those thoughts, my dead alligator returned to life and with a great splash jumped almost completely out of the water as if he had been shot from a cannon. At that point, cool, calm, and collected as usual, I fell on my butt. My waders filled with the cold water I had carefully been trying to avoid, and I started to do the backstroke toward the bank as fast as my water-filled waders would allow. Not fast enough, I might add, to make me feel comfortable. My scientific demeanor and rationale were out the window as my fight-or-flight instincts took over. I wanted out of there even though I knew the alligator was probably as afraid of me as I was of him.

Someone tossed the very rope I had intended to drag the alligator out of the lake with over my shoulder, a perfect toss (this *is* Texas after all), and they scooted me over the top of the water as if I were a skipping stone. Back on land, there was no sign of the alligator.

I was wide-eyed, excited, and a little embarrassed, as well as a bit on the cold side. So I was right after all; there were no flies and there was no odor—the alligator was indeed alive. To the volunteers I said, "See, I told you so" (what an act! he sure looked dead to me as well at the time). Cold, wet, and humbled I returned to the Nature Center to find some warm clothes and soothe my bruised ego.

Sometime later I met the same alligator again. This time he was really near death. He looked poor; he was emaciated and practically skin and bones. I noticed at that time his right foreleg had been broken and mended in such a manner that he could not use it, as it pointed up into the air. I suspect it was from doing battle with one of his peers at some time. Time finally caught up with the old fellow, and he really did die. By the time I found him, he was indeed being visited by flies and was stinky. I covered him up with a big mound of dirt with the intent to recover his skeleton for display sometime in the future. A park visitor found the mound and stole the skull.

Alligators Occupying Unusual Habitats

Sometimes alligators are reported being seen at places out of their range or habitat. If photographed or captured, their presence in these unusual places is validated, but it is usually unknown how they got there. Were they captive animals that escaped or were turned loose? Were they wild animals that somehow ended up in an out-of-the-ordinary place? These four stories are validated cases of such alligators. Three of the accounts involve alligators found in West Texas in the Rio Grande, and the fourth is about an alligator that washed up on a beach.

The Big Bend Alligator

In the 1990s an alligator was frequently seen in the Rio Grande in the Big Bend area by river guides and other people taking rafting trips or canoeing. Most Texas herpetologists who frequented West Texas, including "Doc" Jim Dixon and Mike Forstner, believed that the alligator (if it actually existed) was an escapee from Warren "Gator" Lynch, the first owner of Ring Huggins' Rock Shop at Study Butte (see chapter 7). I got to know Ring, who denied that the allegation was true. I also spoke with an individual involved in rafting trips in the area, and she called that assertion an "urban legend." I might add that Warren felt sorry for the gator being by itself and thought that it should have some companions released with it. No one else agreed, and no alligators were known to have been released for such a purpose. A number of people I talked with seemed to think that the animal was an escapee from the zoo in Ojinaga, Mexico (across the river from Presidio), due to flooding of the exhibit.

River guides and rafters first remember the animal when it was about 5 feet (1.5 m) in length, and they used to see it lying on the grassy banks beside the river on a 3-mile (4.8 km) stretch between Madeira Canyon and Grassy Banks in Big Bend Ranch State Park. It then appeared to move about 9 miles (14.5 km) downstream near Fresno Ranch, which is about 5 to 6 miles (8.0 to 9.7 kg) upstream from Santa Elena Canyon. I also bumped into

Dave Gyure on a trail at Big Bend National Park. He was president of the Texas Herpetological Society in 1968 and a professor at Southwestern Texas State University before retiring. He saw an adult alligator in 2000 while boating upstream from Santa Elena Canyon at Colorado Canyon, which is upstream from Ranch Rios and Madera Canyon. The distance from Ranch Rios to Madera Canyon is 14 miles (22.5 km). The year 2000 seems to be the last time that the gator was spotted. There were photographs taken of the alligator, as well as its footprints. The alligator was reputed to be about 10 feet (3 m) in length at the time of its last sighting. "Catfish," a river guide, probably has the most recent photo, which he took in September 2000. I also heard a single report that a 2-foot (0.6 m) alligator had once been sighted in the same area that the Big Bend alligator had frequented.

Then Mike Forstner, a biology professor at Texas State University, saw tracks in 2005 at Santa Elena Canyon that convinced him that an alligator might really exist at Big Bend. Lisa McDonald (who did the illustrations and took some of the photos for this book) and I went to Santa Elena Canyon in 2005 and saw alligator tracks, and she photographed them (fig. 4.1). Most of the tracks seem to have been washed away by the water, leaving only tracks at the periphery of the wet area (and no tail drags). In 2006, Mike and Bill Zeigler (a zoo herpetologist) looked for tracks or other evidence of the alligator in Santa Elena Canyon and found none.

Mike brought the alligator captured in Laredo (discussed later) to my attention by suggesting that perhaps it could be the Big Bend

alligator. The last evidence that we had from footprints of the Big Bend alligator was in 2005 at Santa Elena Canyon, and Mike did not see any evidence that it was there in 2006. I looked for it in December 2007 when I went back to Santa Elena Canyon. Although it was the wrong time of the year for spotting alligators, the weather was mild. My sister and I saw no alligator tracks. Could the Laredo alligator and the Big Bend alligator be one and the same?

We had no additional information on the alligator at the time and had to deal with the fact that it had to either go over or avoid the Amistad Dam at Del Rio. Also, it would have had to travel much farther than the distances it had moved downstream in the past. Research in Australia indicates that saltwater crocodiles, rather than swim, catch currents in rivers or oceans to move them great distances. It appears that they consciously do this rather than it being a chance event. It is possible that an alli-

Figure 4.1. Alligator tracks in Santa Elena Canyon during a 2005 visit. (Photo by Lisa McDonald)

gator could do the same, perhaps motivated by wanting to find a mate or move to a locale that was more favorable in some way.

I spoke with a river guide from Far Flung Adventures in September 2010. She had heard that in 2008 a Big Bend Ranch employee had sighted an alligator. I talked with "Catfish" about this supposed sighting, and he suggested that I contact David Long at Big Bend Ranch State Park, as he had records of the alligator sightings and was located at the Barton Warnock Environmental Education Center, in close proximity to the river, and talked with the river guides frequently.

Dave sent me a photo of the Big Bend alligator taken on July 9, 1996, where it was first sighted on the far side of the river across from the Teepees roadside rest area on FM 170 about 15 miles (24 km) west of Lajitas (fig. 4.2). Teepees is located 1 mile (1.6 km) upriver from Madera Canyon. The alligator's estimated length was 6 feet (1.5 m). Dave also

Figure 4.2. Photo of Big Bend alligator taken by a visitor to Big Bend Ranch State Park on July 9, 1996. (Photo provided by David Lang, TPWD)

sent me a copy of Mike McDonald's "Outdoors Forum" column from the *San Antonio Express-News*, dated November 1, 1997. A San Antonio resident, Warren Davis, had gone to Big Bend National Park for spring break with his father and brother, and they saw and photographed an alligator while on a Rio Grande rafting trip. The photographs were not high quality but clear enough that Mark McDonald could tell that it was an alligator.

Dave also shared other invaluable information about alligator sightings recorded on National Park Service Natural History Field Observation forms. There were the sightings at the Madera Canyon/Grassy Banks/Fresno areas. The new information was of a 2007 sighting of a 5- to 6-foot alligator in Santa Elena Canyon. The eye-to-nose length of 4 inches (10 cm; which is highly correlated to the alligator being a total length of about 4 feet [1.2 m]) had been sketched on the form with the gator head and those associated measurements. So the alligator from this conflicting information was 4 to 6 feet in total length. The other report of interest, from the Big Bend National Park Bear Database on March 19, 2004, involved boaters, approximately 5 miles upriver from Lajitas, who spotted two large alligators feeding on a dead black bear. A Mexican male came out with an AK-47 and shot at the alligators, which disappeared into the river. A follow-up report stated that there was only one alligator, it was shot dead, and the bear was actually a black calf that had gotten stuck in the mud. The cow makes sense, but the assurance that there was only one alligator was not proven to me. What if two alligators were present: one was killed and the other headed downstream toward Santa Elena Canyon away from the human predator? David Long confirmed the presence of two alligators of different sizes that he saw together and separately. He thought that my idea was quite plausible that the alligators had been together feeding on the cow. At any rate, each time I go to Big Bend, I am going to keep an eye out for the Big Bend alligator.

The Laredo Alligator

In early October 2006, an alligator was seen in the Rio Grande near Laredo, Texas, and Nuevo Laredo, Mexico. It was captured and taken to the Animal Protection Association's shelter in downtown Nuevo Laredo. The alligator was described as being very healthy. It was kept in a dog kennel with a wading pool immediately adjacent to other kennels containing dogs. The total length was determined to be 7.5 feet (2.3 m) with a weight of 130 pounds (59 kg). An article from a Laredo newspaper was written about three days after the animal was captured. The alligator had been tranquilized (presumably to capture it) and was not going to be fed until they were sure that the effects of the tranquilizer had totally worn off. It had been given the name "Cokis," and more than 800 people had come to the shelter to visit it. The plan was to exhibit the alligator, but the article did not detail where its permanent home for exhibition would be. There was only speculation on where the alligator came from. It was thought that someone had turned it loose.

The El Paso Alligator

This is a multiple alligator sighting that has been confirmed. The location is Hudspeth County, just east of Fort Hancock near El Paso. In June 2009 alligators were validated by TPWD game warden Ray Spears. Three alligators were observed that measured between 2 and 4 feet in length, and three were noted to be between 5 and 6 feet in length. The alligators were "left where they were found" but were being monitored. Border fence workers discovered their presence. It is thought that they were "dumped" by owners keeping them captive.

The Padre Island Seashore Alligator

Eddie Sunila, the reptile/aquarium supervisor at the San Antonio Zoo, shared with me that a male alligator named Rufus living at the zoo first showed up on a beach at Padre Island Seashore. Eddie forwarded me an e-mail that he received from Darrell Echols, who at the time Rufus was found worked at Padre Island National Seashore.

According to Darrell, "In 1992, Rufus was brought to the Ranger Station one evening by visitors that had found him washed onto the beach around the 55-mile mark at Padre Island National Seashore. I was the only one working at the Ranger Station, so I visited with them. The visitors stated that there were three alligators that had washed ashore. Two were dead, but Rufus was still alive. They wrapped his mouth with duct tape and transported him to me as they left the park. Rufus was approximately 2½ feet long. We transported Rufus to the permanent pond south of the Ranger Station adjacent to Park Road 22 and released him. He swam into the pond and was observed there for the next several years.

"Periodically, we would find him in the wetland areas in the center of the island west of Park Road 22, but as those wetlands dried up, he would move back to the permanent pond. In 1996, Rufus was observed by contract staff just south of the Bird Island Basin Road near one of three historical oak mottes. He was about 6 feet long at that time. By 1999–2000, Rufus had relocated to the sewage evaporative ponds west of the Visitor Center. We suspect that a more reliable food source, consistent water, and larger area afforded him a better habitat. In 2000–2001, the park reconstructed these evaporative ponds and Rufus moved to the wetland areas in the center of the island. About six to eight months after the reconstruction ended, Rufus came back to the pond, where he stayed until we relocated him to the San Antonio Zoo.

"Normally the National Park Service does not relocate animals from parks, but Rufus had become habituated to humans. For three to four years, park maintenance staff fed Rufus anything, from roadkill, leftovers, hot dogs, and more despite repeated chastising from the superintendent and me. Rufus eventually expected a handout from whoever visited the sewage ponds, which posed a safety risk to both park staff and Rufus. I decided that we had no choice but to remove Rufus from the park. We examined other areas suitable for him, such as elsewhere in the park, Aransas Wildlife Refuge, several state parks, and Welder Wildlife Refuge, but these areas were also opened to visitors and in essence just transferred the safety issue

to another area. We considered killing Rufus, but I indicated that this would be a last choice after we had exhausted all other possibilities. We contacted the San Antonio Zoo, who happened to be looking for a large male alligator since theirs had died, and the rest is history. We were very thankful that the San Antonio Zoo had the desire and option to take a large male alligator. I have a photo of us removing Rufus hanging on my wall next to my desk, and many staff and visitors coming to my office ask about what we were doing."

Eddie reports that Rufus has been at the zoo ever since, living life with a female alligator.

Saving Alligators in Distress

Hurricane Ike (September 13, 2008) impacted wildlife as well as human inhabitants of the upper Texas Gulf Coast. The following two stories involve Gary Saurage, co-owner of Gator Country Alligator Theme Park (see chapter 7) in Beaumont, Texas, and his before-and-after Ike activities involving his captive animals, as well as his post-Ike rescue of wild alligators.

Preparing for the Storm

Most business owners board up their businesses and leave prior to a hurricane. When you are in an animal business such as a zoo, wildlife theme park, or private collection, it is quite a different situation. The ideal solution is to evacuate the animals. When that is not an option, the responsible person stays behind, makes conditions as safe as possible for the animals, rides out the storm, and deals with the aftermath in terms of the animals' well-being.

Gary Saurage had nearly 300 alligators, additional crocodilian species, snakes, and other reptiles at Gator Country. Evacuation was not an option. There were 50 alligators in an area enclosed by wooden fences in front of the restaurant associated with the reptile theme park. Wooden fences do not last in a hurricane, so the 50 alligators had to be caught by hand and moved to a pond that had hurricane fencing. The pond for the crocodiles had to be drained in case Taylor Bayou overflowed its banks and flooded the nearby exhibit. Generators had to be ready to go online at the snake exhibits to prepare for power outages. The outdoor buildings and the baby alligator incubator house had to be secured.

During the hurricane, everyone who stayed behind to care for the animals was stationed at the restaurant. The park was scanned with a large spotlight every few minutes to see what was transpiring outside and to look for trees falling over fences. Each time trees in the park were seen going over fences, Saurage and the others staying behind to take care of the animals ran out in 85 mph winds to cut them off. The situation was made even more dangerous as the tin roofs from the pavilions were torn off and were flying through the air. The individuals who went out during the hurricane to tend to the fences had to crawl alongside the hurricane fencing to keep the flying debris from hitting them.

After the hurricane, the animals outside needed to be checked to make sure that they were alive and well, in addition to surveying the integrity of their enclosures. "Big Al," the large male alligator that was one of the few survivors of the original alligator farm that became Gator Country, was not seen for several hours after the storm. Most of the trees near the pond in his enclosure had been cut down just prior to Hurricane Rita as a precaution. During Ike, the winds came from the opposite direction, and some really big oak trees fell into his pond. For several hours, Saurage saw no sign of his old friend and worried that a tree had fallen on top of him and he had drowned. At daybreak, Saurage entered Big Al's enclosure and splashed the water at his pond as a signal to "call him." Big Al's head emerged from the water, and Saurage gave him a tap on the nose before moving on to check out things at the rest of the park.

Rescuing Alligator Survivors of Ike

Gary Saurage drove down Highway 73 in Jefferson County after Hurricane Ike ravaged the upper Texas Gulf Coast to survey the impact on the alligators and other wildlife. He saw at least 150 dead alligators visible from the road. He knew that the toll in the marshes and rice fields, unseen from the roadside, must be catastrophic as well. These lung-breathing animals had been drowned or near-drowned by the high water and the salt water. Saurage has a nuisance alligator hunting permit for the state of Texas, which would prove to be invaluable for the rescue that he wanted to make for these alligators. A couple of days after the hurricane, he met with a state representative and discussed how Ike had devastated the alligator population in the area. The representative asked if he had the resources to rescue at least some of the alligators along Highway 73. He said yes. State Representatives Mike Hamilton and Alan Ritter paved the way for him to get the alligators rescued, and Monique Slaughter of the TPWD Alligator Program gave him the go-ahead to save as many alligators as he could. They readied an enclosed one-acre pond for housing the alligators and rescued more than 40, which were roped or, in most cases, jumped to subdue them. The conditions for capture were horrendous, as the water was contaminated. It contained oil, botulism, and flesh-eating bacteria. Most of the alligators had eye infections. Due to the oil, it was difficult to tape their mouths shut with electrical tape. They had to use sandpaper to "wipe" the oil off their mouths to make the tape stick. A couple of times Saurage and his staff incurred bites to the arm. The captured alligators spent three weeks at Gator Country, after which they were released back into the wild.

Neighborhood Alligators

This account is told by Bill Paske, an engineer who has been a fellow board member with me for the Rice University/Medical Center Chapter of Sigma Xi, the Scientific Research Society. "I live in a subdivision south of Houston, which has a nice tennis court placed next to a rather large drainage ditch. We have had many occasions where a small alligator would come up on the sidewalk or even the court itself to warm up. One rather dark evening, another homeowner and I were working on the lighting for the courts, which did not want to work correctly. We were stretched out on the grass, grunting and reaching into the power well

trying to sort out the wiring when we heard a different grunting sound. I did not see anyone in the dim light when I looked up, so we went on about our business. When I heard the grunting again, we turned the light in the direction of the grunting and saw what appeared to be a 12-foot alligator moving in our direction. He stopped when we stood up. We left the power well with the cover open and decided to retreat to our car. The alligator moved back to the ditch, and we each chided the other as the object of the alligator's intentions. I guess lying on the ground in the dark, with our grunting and trying to move stuff in the well, we appeared interesting, enticing to him. When he left, we replaced the cover and finished the work the next morning when we could see who was coming to visit us."

Research Design by Innovative Students

This story is also by Bill Paske. One never knows how research projects get designed. This gives some insight. I had read the publication that resulted from this research when I first started studying alligators in 1985, about 10 years before I met Bill. It is amazing that what started out as a joking comment by Bill was taken seriously by a student and resulted in a scientific publication.

"I had a graduate student friend whose major was biology. She was interested in studying catfish as they moved around their tank in her lab. I gave her some pointers on how to build a telemetry system so that she could monitor their temperature and pulse rate. She built a system that worked on her own. Later that semester, she brought a friend over to a party. After a while, he began to discuss his project: studying the thermal properties of alligators. We don't have many alligators in Oklahoma, so he wanted to be able to calculate heat transfer rates and compare these rates to real measurements for known sizes and then extrapolate to other, larger alligators. I was feeling rather flippant and suggested that he assume cylindrical alligator geometry and then apply the known heat transfer functions to this rather simple geometry. This was meant partly as a joke, as we had been telling jokes about experimentalists and theoretical physicists and how they perceive the world (e.g., a spherical cow). However, the student was bright and innovative and soon had a first-order model, which actually provided a reasonable result. We then had a serious discussion as to how one would combine different geometries to more closely approximate a real alligator."

Observatory Alligators

A book by Timothy Ferris, *Seeing in the Dark: How Backyard Stargazers Are Probing Deep Space and Guarding Earth from Interplanetary Peril* (2003), includes an interview with Barbara Wilson, staff astronomer and assistant manager of the George Observatory at Brazos Bend State Park. The observatory is located across the road from the Interpretive Center and its adjacent parking lot. The path to the observatory bisects Creekfield Lake with a bridge that was constructed to facilitate access to the observatory when it was built. The interesting thing about this interview is that it starts off with a discussion about alligators, not about stargazing. Does this mean that alligators are more attention grabbing than stars and planets? Probably so.

The discussion centered around the fact that alligators would not be a problem for Barbara, but they would discourage groups of schoolchildren from crossing the bridge over Creekfield Lake to access the observatory when they had a scheduled tour. If a group was late, Barbara would suspect that perhaps they were being held up at the bridge by an alligator basking. She would take a rake and go forth to remedy the problem by scraping it in front of the alligator, and the movement and sounds would generally convince the alligator to take off. She commented that the kids think that it is cool, but the adults don't agree, especially the men. The comment was made: "What is it about men and alligators? A six foot alligator will show up and the men won't do a thing about it. Instead they'll call me on their cell phones and wait for me to come over with my rake. The other day there were five or six grown men just standing there, looking at the alligator and asking me to deal with it. So I chased him off into the woods." She notes that the bigger alligators aren't much of a problem, since they're more experienced and have some fear of humans. She says that the little ones will try to snap at you when you shoo them away, trying to turn around and come back. She also comments that it is hard to herd alligators and that they are very gentle with their young.

Two Very Short Stories from Perry Smith

Perry Smith of TPWD was one of the people who helped me undertake my alligator research at the J. D. Murphree WMA. He usually drove the airboat while I conducted my mark-recapture study of juvenile alligators. He gave me sage advice in regard to working in the marsh or with alligators. For example, he told me not to eat bananas before going out in the marsh or I would get eaten up by mosquitoes. One of the items on my list to buy for research was hip waders. He cautioned me to never use them in the marsh because I might stumble into a gator hole and sink. I heeded his advice and never bought those hip waders. A couple of times I was wearing regular rubber boots that came up to just below the knee, and I stepped into an alligator hole and had a difficult time extracting myself from the water that weighed me down. I hate to think what would have happened had I been held hostage in those waders, and thank heavens those were not occasions in which I had alligators come after me! Thank you, Perry, for taking such good care of me!

The following short descriptions of alligator behavior show just how quick alligators can be: Perry was in a public place and noosed a nuisance alligator. There were kids all around him, and the alligator shot straight up into the air. Needless to say, it made quite an impact on the kids, who scattered. On another occasion, Perry was releasing some banded ducks into the wild. As he released one duck, an alligator grabbed it in midair.

Alligator Hunting and Farming

Wealthy and fashionable ladies clutching alligator bags have seen the alligator as the source of the sensual pleasures of very fine and exotic leather. Poverty and deprivation in the past have led Americans to eat the alligator for want of other foodstuffs; while today, alligator meat has become an expensive gourmet delicacy with worldwide distribution.

—Vaughn L. Glasgow, *A Social History of the American Alligator*

The alligator is potentially a very important resource to Texas as an economic commodity for hunting, farming operations, and tourism. The alligator populations throughout the United States declined due to overhunting but rebounded after having been afforded protection over their entire range by the US Endangered Species Act of 1973. By the mid-1980s, the population of alligators in Texas was believed to be over 100,000 and possibly even several times that number. The general consensus was that alligator numbers in the United States were not quite as low as had been estimated at the time that legal hunting ceased. It was thought that the alligators were wary in the presence of humans, and, in their effort for survival, low visibility was a key survival tactic. Therefore, their numbers were underestimated when they were at their lowest level.

As discussed in chapter 3, alligators are long-lived animals that are slow to mature. Each breeding-age female does not reproduce every year and is capable of producing only one clutch of eggs the years that she does reproduce. If the eggs are not predated and successfully hatch, there may be more than 40 young alligators that result from a nest. However, the high predation rate on young alligators and a substantial mortality rate for subadults mean that only a small percentage of young alligators live long enough to enter the breeding population. Hence, there was a substantial lag time before alligator populations could recover to levels where hunting was again feasible in terms of a sustained yield harvest. This means that the hunting level over a period of years will not cause the population to decrease.

In order for a population to be harvested on a sustained yield basis, the population needs to be managed or controlled. A harvest cannot be haphazard. Criteria must be set concerning which localities have a population level high enough to allow hunting and what percentage/ numbers of animals will be harvested. Moreover, the harvested populations need to be monitored over time concerning population surveys, reproductive biology, habitat selection and quality, and other aspects of population dynamics. Once a population is harvested and managed, it is no longer a naturally occurring population. Rather, it is a managed population that occurs in the wild. It is subjected to having its numbers, size classes, sex ratio, and social interactions altered by the management strategies; various environmental factors and chance occurrences also come into play. Therefore, it is important that some populations not be harvested in order to have the capability to compare them to the managed populations.

Webb and Smith (1987) noted that a good site for studies on crocodilians should have accessibility, no harvest, a history of past hunting, a sizable population that is not in marginal habitat, good long-term study prospects of the site, available nearby weather station data,

aerial photographs for the site and vegetation data, and known general distribution patterns of animals and nests. In Texas, Brazos Bend State Park fits many of these characteristics and is the best alligator study site. Additionally, it is ideal because it can be accessed on foot without the use of an airboat, and it is not a continuous coastal marsh but has smaller water bodies that are easily delineated.

Resuming Alligator Hunting in Texas

The status of the alligator in Texas was reclassified from Endangered or Threatened to Threatened due to "Similarity of Appearance" by the US Fish and Wildlife Service on November 14, 1983. The term "Similarity of Appearance" refers to a species that in its entirety or "parts thereof" resembles another species that is Endangered or Threatened. There is a concern in such cases that the species whose numbers are critically low may be "passed off" knowingly or unknowingly as a species that is more common and is legal to harvest.

In 1984, the first legal harvest of the alligator since its protection was held September 7–23 in both coastal and inland habitats. There was considerable fanfare associated with the hunt. McIlhenny's *Alligator's Life History*, which had been reprinted in paperback in 1987, was placed in multiple copies on the shelves of libraries in the areas to be hunted. Channel 8, the local Houston PBS television station, put together a program on alligators and the hunt. I was interviewed about alligators, since I was starting my doctoral research on them. I remember going to the Houston Zoo (where I worked prior to starting my doctorate) for the filming to "borrow" one of the juvenile alligators that I had formerly cared for there to use as a prop. The program included footage from the beginning weekend of the hunt, and it was shown during that first hunting season, as well as off and on for a number of years afterward.

TPWD made an informational video on hunting and skinning alligators for the hunters that first year. A biologist from TPWD spoke to the Wildlife and Fisheries Sciences students at Texas A&M in College Station about the upcoming hunt and showed the video. It was said more than once in the video to "make sure the alligator is dead." All of us were laughing really hard, as it seemed incredibly funny that the comment was interjected within the video repeatedly.

I accompanied Bruce Thompson, the Alligator Program leader for TPWD at that time, to survey and monitor the hunt in the coastal counties that first weekend. The J. D. Murphree WMA at Port Arthur was the staging area for hunters to get information and to bring alligators in to a check station set up there and was also a location for a limited number of hunters to hunt alligators. Hunters were in the office area at Murphree WMA watching the video that I had seen previously at Texas A&M. They had their eyes glued to the screen and did not laugh at the "make sure the alligator is dead" remarks. Everyone involved with the hunt, the TPWD officials, hunters, landowners, and locals hired to skin or process alligators for commercial operations, wanted things to go smoothly and were extremely cautious

about doing everything according to the letter of the law.

Setting Harvest Limits

Just how does TPWD decide how many alligators to hunt; or, to put it another way, how do they count alligators? Surveys of the alligator population are conducted each year via aerial nest surveys and spotlight counts to determine how many tags would be issued. Nest surveys are conducted after the alligator nests have all been built. A leased helicopter flies over established transects with someone from the TPWD Alligator Program marking the locations of nests on a map. I was invited to be an observer for one of the survey routes in the mid-1980s. The helicopter was a two-seater and had a bubble-type canopy window and no doors, which maximized visibility. Monique Slaughter of the TPWD Alligator Program and I were sharing a single seat belt, as well as a seat. I feared that one of us would accidentally unbuckle the seat belt and we would both fall from the sky into the marsh, but I learned how to spot an alligator nest from the air. The only other structure that could be mistaken for a nest is a muskrat lodge. Muskrat lodges are entirely brown, and alligator nests have some green vegetation that was still apparent at the time of the flyover.

Aerial nest surveys are used primarily in the large coastal marshes in Jefferson, Chambers, Orange, and Galveston Counties (known as the Primary Harvest Area by TPWD) and are utilized to some extent in the TPWD-delineated Secondary and Tertiary Harvest Areas. Flyovers can be utilized only in this open type of habitat, as tree cover hides nests in forested areas. About 65 % of the wild harvest alligator tags issued are based on helicopter nest count data. Since nesting can vary from year to year, a three-year average nest count is used in calculations. This is the tag issuance formula:

$$\text{average 3-year nest count} \times \text{nest multiplier (63.5)} \times \text{\% of adult alligators in a population (0.134)} \times \text{harvest rate (0.08)} = \text{number of tags}$$

The formula is based on a 1980 size-class frequency population model that Dave Taylor developed while working for the Louisiana Department of Wildlife and Fisheries (Taylor n.d.; Taylor and Neal 1984). The nest multiplier was one of the calculations generated from the model to estimate the minimum population of alligators per area surveyed. It can be defined as the after-hatching population of alligators/the number of nesting females per 100 adult alligators. It is assuming a 50:50 sex ratio and that 68 % of females nest in a given year. The total after-hatching population of alligators is calculated to be 2,158 individuals per 100 adults. If 50 of 100 adult alligators are females, and 68 % of them nest per year, then there will be 34 nesting females (or 34 nests) in a given year. So the nest multiplier = 2,158/34 = 63.5. The minimum population of alligators = the number of nests × the nest multiplier (63.5). The number of adult alligators is estimated to be 13.4 % of the population, so multiplying the minimum population of alligators by 0.134 yields the minimum population of

adults. Taylor came up with 8 % as the sustainable yield harvest rate, so 0.08 × the minimum population of adults gives the number of adults that can be harvested per year.

The harvest rate started out at a conservative 4 % the first year and was changed to an 8 % rate thereafter. The population model that I described in my research indicated that 8 % may be too high for a sustained yield if the sex ratio is not 50:50 (i.e., there are more females than males in the population). However, the tags available to be issued are always greater than the number actually issued, and the number issued is always higher than the number used (alligators that are harvested) overall. The total number of tags issued per year by the State of Texas also includes those used for nuisance alligators. It is also important to note that calculations of how many tags are issued each year are estimates of how many alligators can be hunted, arrived at by using estimates of minimum population levels, not actual levels (which are hard to determine). Thus, the numbers are not 100 % accurate. So if an actual 8 % harvest rate is truly too high, Texas seems to be harvesting below that level.

Spotlight counts and harvest trends are used to determine the tag issuance rates for the approximately 35 % remainder. In spotlight counts, an airboat is utilized to navigate the waterways and a high-powered light is used to catch the red glow of alligator eyes at night. Each pair of eyes is counted, and whether the alligator is greater or less than 6 feet (1.8 m) is recorded. Animals whose length is unknown are described as such. Set transects are used to compare population levels over a period of years.

Licenses, Tags, and Associated Paperwork
A separate alligator license used to be required in addition to a hunting license. Long-standing complaints that hunters were being charged twice have come to an end, and now alligators may be taken under any resident or nonresident hunting license.

A hide tag must be attached near the end of the tail just below the fins. The hide tag is not a state of Texas or federal tag but a CITES tag. CITES is an international agreement among governments to ensure that international trade in specimens of wild animals and plants (both living and dead) does not threaten their survival. The United States and other countries pass their own domestic legislation in the form of a licensing system to ensure that CITES is implemented in their particular country.

A landowner must apply through a TPWD Application and Receipt for Alligator Hide Tags to get the tags, and any tags that are not used must be returned to TPWD. After the hunt, the landowner must submit a TPWD Alligator Hide Tag Report that lists the individual hide tag numbers in association with the hunter who killed a particular alligator and his or her license number. It also lists the tags that were not used and categorizes them as being returned intact, damaged, or lost.

There is also an Individual Hide Tag Report to be submitted and signed by the hunter that requires this specific information: landowner/agent recipient number; county where harvested; the hunter's name/contact information/license number; the hide tag number; sex of the alligator; length of the carcass; if hide use is commercial or personal; if the

skinning method is the belly skin or hornback (mount); and if the alligator was dispatched (killed) with a shotgun, archery, or other means. The hunter submits a copy to the TPWD Alligator Program at Port Arthur and keeps a copy. Another copy stays with whoever has custody of the hide until it is shipped or sold out of state, and then the copy is returned to Port Arthur. In addition to the Individual Hide Tag Report, an Alligator Parts Label report must be completed that encompasses "parts thereof," such as meat, teeth, claws, and bone/skull.

At the time of the first hunt, a hide flap was required for an alligator hide to be legal, in addition to the tags and paperwork. A hide flap is a rectangular portion of skin cut from a specified section of the alligator's back and left attached to the rest of the hide. If a different section/side of the body is used for a hide flap each year, it is possible to ascertain what year the hide was harvested. It serves as a backup to the hide tag and helps ensure that the tag and the hide go to the same animal that was harvested. I recall that for the first hunt the location of the hide tag was just in front of the hind feet and was to be taken from the right side. The media were filming a man skinning an alligator and asking him questions. He was distracted and ended up cutting the hide flap on the opposite side. When he realized what he had done, he backed away from the alligator and called over TPWD officials who confiscated the alligator. The skinning of the alligator was finished by several of us working at the check station. It was hard work. The following year the hide flap was taken from the opposite side. The process ended up being discontinued altogether, as it was determined not to be that useful in the compliance process. I have often thought about the man losing the alligator that first hunting season and how sorry I felt for him. Things have relaxed a lot since that first hunt and have gone smoothly due to the diligence of TPWD in setting up procedures carefully from the start.

Opportunities for Hunting

Landowners (or their designated agents) who have alligators on their property are issued tags for hunting alligators based on the acreage and type of habitat owned. For example, in 2002, issuance rates varied from one tag per 126 acres (publicly managed leveed freshwater marsh and Keith Lake estuarine system) to one tag per 515 acres (freshwater marsh) in Chambers, Galveston, Jefferson, and Orange Counties. A few landowners provide private alligator hunts to the public.

Hunts on public lands are also available to a limited number of hunters for a fee on state lands such as J. D. Murphree, Mad Island, Guadalupe Delta, James E. Daughtrey, and Angelina-Neches/Dam B WMAs. An additional tagging fee is charged if the hunter decides to sell the alligator. In recent years, youth hunts (16 years of age or younger) on state lands have been held. There are no fees charged unless the hunter opts for a commercial sale of the alligator. All youth hunters must be under the direct supervision of an adult. Youths 12 to 16 years of age who have completed Hunter Education Certification must be under the supervision of an adult but do not have to remain in the immediate presence of the adult.

Commercial versus Noncommercial Hunters

Commercial hunters sell the alligators that they hunt. Some individuals want the experience of hunting an alligator but cannot afford to have high-end leather products made from the hide and eat meat that would fetch a high price at a restaurant. Repeat hunters may keep the meat and sell the hide (just how many alligator wallets, belts, and boots does one need?) or sell the alligator in its entirety (skinning is just too much work). Others are landowners (or work for landowners) who are trying to make at least a partial living off the natural resources on their property. The percentage of commercial versus noncommercial numbers of alligators (based on hides) taken varies from year to year, but thus far there have always been more hides sold than kept for personal use.

There are two mind-sets concerning the size of alligator that is desirable to harvest for noncommercial use. The trophy hunter group wants the largest animal possible, especially if they want a mounted head or full body mount. The other group wants meat and a large-enough skin to get a pair of boots made. Large alligators are very heavy and difficult to drag and lift distances in swampy terrain. For this reason, I have heard hunters who caught very large alligators say that they would rather take smaller animals in future hunts. The smallest size that is legal to hunt is 4 feet (1.2 m), but the animals less than 6 feet (1.8 m) are usually not too popular with either commercial or noncommercial hunters. They yield less meat and leather and are not as impressive as a larger animal. However, the smaller alligators are the group that is the most sustainable to hunt. It takes many years for an alligator to become really large, and while the numbers of alligators in a hunted population may remain stable over time, the number of very large alligators typically decreases. There are only so many of them out there, and if a population is hunted over time, it is likely that more large animals will be taken than can be replaced by the growth of smaller individuals.

Legal Means and Methods to Harvest Alligators

The first constraints on hunting regard lawful hunting hours. "From one-half hour before sunrise to sunset. Between sunset and one-half hour before sunrise, no person shall set any baited line capable of taking an alligator, remove an alligator from a line set, or use any means and methods other than line sets." These TPWD regulations mean that any legal means to harvest an alligator cannot take place at night. The baited lines known as line sets can remain in place overnight, but they cannot be set or have an alligator removed from them at night. Florida is the only state that does allow hunting after dark at present.

The line sets (hook and line) are the conventional method used to take alligators during the harvest (fig. 5.1). There is a hook to which a baited line is attached. The bait may be a natural prey item like frogs or a human food item like chicken. The hook is a heavy-duty hook, such as a shark hook. The line must be a minimum 300-pound (136 kg) test line per TPWD specifications. The late Bill Howell hunted alligators beginning in 1995 on land he leased near his home in Needville. He noted that the length of line is important as well and used a 50-foot (15 m) line. He said that if it is too short, the alligators may bite at it but may not get the hook inside their mouths (so they get away). Ideally, they should be able to grab the bait with the hook and run away with it but get caught up by the hook. A longer line also helps keep them safe from being attacked by a larger alligator. (There have been a number of accounts in Texas, including one at Pilant Lake outside the limits of Brazos Bend State Park, of an alligator that was relatively small being eaten by a larger alligator while on a line set.) The underlying idea is that the alligator is not in a predicament like a dog on a chain that is trapped and helpless. It is still able to move around and hopefully be able to thermoregulate as well. Basically, the goal is for it to be "comfortable" until it is dispatched.

The end of the line that contains the hook and bait is attached to a stationary object such as a bamboolike sea cane pole that is stuck into the substrate (bottom) of the water body. The baited hook is placed on the pole high above the water to prevent smaller alligators from taking the bait. Large alligators have the ability to leap several feet out of the water, so this measure helps ensure that only the larger animals are hooked on the line. Using a wooden dowel that is about 3/4 inch (1.9 cm) in diameter and about 4 inches (10 cm) in length instead of a hook is also a means to avoid hunting an alligator that is too small. The dowel has a hole in the center to which the line is attached. If a small alligator grabs a baited wooden dowel, then the line can be cut and the dowel will pass out the digestive tract without harming the animal.

a

Figure 5.1. (a) A baited line set. The bait is held well above the water to exclude smaller alligators from being able to reach the bait. The line set is the most common means used in hunting alligators. (b) Bill Howell pulls an alligator to shore that has swallowed the baited hook. A shotgun or bow and arrow are then used to kill the alligator. This alligator was lucky, as Bill cut the line and let it go free as it was a bit on the small side.

b

The other end of the line is secured to a tree, wooden post, or other strong object capable of securing the line after the alligator grabs the bait and until the hunter retrieves it. A tag must be attached to this end of the line that contains the name and current address of the hunter, the hunting license number, and the hide tag number. This helps ensure that lines are legally set and that the hunter can be tracked down if he or she is not in compliance with some aspect of the regulations regarding alligator hunting.

An alligator on a line set can be killed with a firearm or by a barbed arrow. The advantage of the arrow is that it causes considerably less damage than a firearm. At the time of the first hunt, I wanted a skull, and Don Broussard agreed to fill out a parts label for a "relatively intact" head I found after the animal was dispatched by gunfire in his commercial operation. The skull that I chose did involve some shattering to the extreme right mandible and to the left frontal region (a diagonal shot) but was relatively intact. Most skulls sustain considerable damage from gunfire. For this reason, hunters who want to keep the head and skull as intact as possible use a bow and arrow to dispatch an alligator.

The other less common means to take an alligator are an alligator gig (a spearlike device with a long, thick handle and immovable prongs or two or more spring-loaded grasping arms), a hand-held snare with a locking mechanism, archery equipment with a barbed arrow, and center-fire firearms (can only be over private property/water). In my opinion, the most humane method of hunting an alliga-

tor is via the line set (hook and line method), followed by killing it with a gunshot or arrow to the skull in the region of the brain. Additionally, any method needs to be followed up with destroying the brain by pithing it with a metal rod if there is any chance that the brain is not totally destroyed, which will prevent the animal from suffering. As discussed in chapter 6, TPWD's own experimental hunt with firearms showed it to be an ineffective method for hunting and retrieving alligators (Johnson and Thompson 1986). If an alligator is hit, it may die a slow, painful death or become dangerous. Many biologists besides myself, including some employed by TPWD, feel that it was a big mistake to change the regulations and allow alligators to be hunted on private property with firearms.

Skinning an Alligator and Preparing the Hide

The media interviewed hunters at the Murphree WMA during that first hunt as the alligators were brought in to be checked. Some individuals were skinning alligators right there, where setups for that purpose were available. If a taxidermy mount is to be made from the alligator, the entire carcass is skinned, and the skin is known as a hornback due to the presence of the osteoderms embedded in the skin of the back. Typically the skin taken from an alligator is the soft belly skin from the underside of the animal from the head to the tail and the skin from the sides of the body. The leg skin is removed to the "wrist" or "ankle" area. The portion of the tail that has a single row of fins at the top is included in its entirety to the tip of the tail. The cut is made just below the base of the fins. It is important that the tip of the tail is included as the value of the hide is largely determined by its length. The heavily armored skin of the back with the osteoderms embedded in it is left on the alligator carcass. It is important that the hide is carefully cut from the body and that no holes are poked in the hide, which would adversely affect its value.

Large alligators are more time consuming to skin, not only because of their size but because both pulling and cutting are required to remove the belly skin. On smaller alligators, after the time-consuming cuts are made and only the actual belly portion of the skin remains to be removed from the carcass, the skin can be removed by pulling down only to the area of the cloacal opening or vent. The skin has to be cut carefully away from the cloacal opening, and then the skin of the tail is pulled and cut until the hide is separated from the carcass. Any remaining meat or fat must carefully be scraped from the hide so that it does not rot. Although alligators do not have to be "field dressed," it is important that the hide not be allowed to dry out or be exposed to the sun before, during, or after the skinning process to avoid scale slippage, which damages the hide.

The hide is washed with clean water and hung on a rack to dry. Then the hide is liberally salted all over with fine-grained salt that is thoroughly rubbed in. The final step is to carefully roll the hide into a compact bundle with the flesh side to the inside. This usually requires two people to keep both sides even and the roll as compact as possible. The hide is tied with cotton string, and the cotton string is also run through the metal hide tag that is attached near the end of the tail. This step ensures that the tag stays with the rolled hide. The skin is stored in a cool, dry place until it is sold or processed. The storage area must provide adequate drainage, as the hide will drain. If a hide is stored more than a few days, then it is necessary to unroll, resalt it, and then roll it again. After this step it is possible to keep hides several months if stored properly.

Timing of the Hunting Season

Since the restart of hunting in 1984, the alligator hunting season has always been timed to be no earlier than the first week of September in order to avoid the mating and nesting season. It is also prior to alligators going off food for the colder months and becoming less active. The season has been standardized to September 10–30. Most of the hunting takes place during the first weekend, when Gatorfest is held.

The Controversial "Second" Hunting Season

It is unheard of to hunt an animal resource during its reproductive cycle, yet that is what TPWD is doing with the American alligator during the period that mating occurs and the females are nesting. This hunting season first started April 1, 2007. It does not include the 22 core counties of Angelina, Brazoria, Calhoun, Chambers, Galveston, Hardin, Jackson, Jasper, Jefferson, Liberty, Matagorda, Nacogdoches, Newton, Orange, Polk, Refugio, Sabine, San Augustine, San Jacinto, Trinity, Tyler, and Victoria. These southeastern core counties are where most of the state's alligators are located

What Was the Fate of This Mother Alligator?

On April 22, 2007, I received an e-mail from a concerned citizen about a female alligator with babies in the Clear Lake area (near Houston) that inhabited an area in a bayou off a golf course in the Bay Oaks subdivision. The alligator had been seen with as many as 18 young, but the number had dwindled to 9. Several people were believed to have captured some of her offspring. She had rocks and golf balls thrown at her, and one teenager shot her in the eye with a bow and arrow. Some individuals were thought to be feeding her as well. Not a cool situation . . .

The individual who contacted me went to the TPWD website and read the "Do's and Don'ts of Living with Alligators" and posted this information on-site in the form of huge signs to discourage people from harassing the alligator and her young and to stop people from thinking that they were helping by feeding her. This individual was interviewed on a local radio station to discuss the alligator and her plight. Other people supported this alligator and her pod as well. She was named "Alice," and they got her a spot on Houston television Channel 11 news.

My contact informed TPWD about the entire situation with the alligator. She was assured that the abuse that the alligator had suffered was against the law, and if the name of the child who had shot her with the bow and arrow was given to them, they would deal with it. She was told that if the child was 14 or under, he would not be prosecuted, but his parents would. The child was identified, as not only shooting her but having taken one of her babies. When the informant talked to the game warden to give him the name of the child, he said, "If I hear one more word about this alligator, I am coming out to shoot her!"

I said that I would help to see if there was a way that we could deter her from coming to the spot where she had contact with people. Before I could get further involved, I got a report that the alligator and her offspring had disappeared. We never found out if she was removed by that game warden, a nuisance alligator hunter, or left the area on her own. This was just at the time that the first spring hunt went into effect, and I suspect that she was removed.

The said game warden was not upholding the laws about harassing alligators or possessing them (taking the young alligators). In addition to not doing his job, he made threats about what he would do if he was bothered again. He needed to be disciplined or even fired for this action. Renegade civil servants are not serving the public who pay their salaries. Such individuals need to conform to what is considered professional behavior or find a job in the private sector (where customers can complain about them). And what about letting the juvenile delinquent go scot-free for shooting the alligator and possessing one of her offspring? It is also well known that children who physically abuse animals are ticking time bombs in regard to hurting other human beings.

and where prime habitat exists. Only the September hunting season is legal for them.

Hunting for this April–June season is legal in any of the other 232 Texas counties. Never mind that alligators are not found in most of these counties. So what is the point of having this season? Noncore southeastern counties, such as Harris and Fort Bend (near Houston), and the area of the Rio Grande Valley are having an increasing number of nuisance alligator calls as alligators are moving into newly created habitats associated with decorative ponds, detention ponds, and irrigation waterways associated with new areas of dense human population. Game wardens spend much of their time responding to these calls, which occur most frequently during the breeding season and early nesting season. The goal is to try to eliminate alligators from the areas where they are not wanted, lighten the load on the game wardens, and provide additional hunting opportunities.

Only one alligator per hunter can be taken, and the alligator must be on private property. No alligators may be hunted during this season if any alligators were taken on the same property during the September season. Unlike the September hunt, it is not necessary to be in possession of a CITES tag before hunting. After an alligator is killed, a Wildlife Resource Document (WRD) must be completed right away. Within 72 hours, an alligator Hide Tag Report Form must be completed and mailed to TPWD, along with a hide tag fee in order to receive a CITES tag. The CITES tag is attached as soon as it is received.

Also, some "special properties" in noncore counties may have core county regulations in effect. They are eligible to have TPWD surveys done of the habitat and alligator populations in order to have hide tags issued as they are in core counties. For example, noncore Fort Bend County contains some prime alligator habitat near the Brazoria County line, such as Brazos Bend State Park, and nearby private properties that have substantial aquatic areas with alligators.

Aside from some breeding females and adult males, the alligator "type" that most commonly inhabits new areas in noncore counties is the subadult (3 to less than 6 feet) seeking to establish territory of its own. It is trying to keep from being eaten by larger animals and to find a place to call "home." Remember, the length of subadult alligators that can legally be harvested (at least 4 feet) is the size that most hunters feel is "too small." All things considered, just what was TPWD thinking when it created "the second season"? Admittedly, game wardens should not be spending large chunks of time taking care of nuisance alligators or alligator incidents. It would seem to make more sense to let nuisance alligator hunters or farmers take females and young alligators to farms and harvest the males.

Harvest Data

There were 474 available tags and 437 alligators harvested in 1984 when the first hunt occurred. From 1984 through 2002, there was a peak in the numbers of alligators harvested in 1989 and 1990 (1,956 and 1,955, respectively) with 2,162 and 2,075 tags available for those years. For the same time period, the most tags were available in 1994 (2,209), but only 1,208 alligators were harvested. From 1991 to 1993, a decline in alligators harvested was due to a decrease in the price paid for hides. The market stabilized and then began to climb again in the late 1990s. In 2002, there were 2,182 tags available and 1,869 alligators were harvested.

Products made from crocodilian skins are considered to be luxury items. When the market for the finished goods decreases (usually due to a poor economy), then the price paid for hides drops. Personal use of hides can be expected to increase when the commercial market declines. For example, a high ratio of personal to commercial use for hides occurred in 1999 when there was a nearly equal number of commercial sales (534) and those kept for personal use (505). The year 1989 had a large ratio of commercial versus personal use with 1,745 hides sold commercially and only 185 kept for personal use.

The sex ratio of alligators taken on hunts in Texas has varied from a 3:1 ratio of males to females in the early years of the hunt to roughly a 3:2 ratio of males to females from 1989 through 2001 (although there was a spike in favor of males in 1996 and 2000, giving a nearly 7:3 ratio). In 2002, the ratio became nearly 1:1 (55:45). The average length of alligators taken on the hunt has decreased by about a foot for both males and all of the alligators harvested (males, females, sex unknown). Males had an average length of 9 feet, 1 inch (about 2.7 m) through 1986 and then decreased below 9 feet (or even 8 feet) thereafter. In 2002, the average length of males harvested was 8 feet, 0 inches (2.4 m). The average size of all alligators harvested peaked in 1986 at 8 feet, 6 inches

(2.6 m) and averaged below 8 feet from 1991 to 2002. From 1997 to 2002, the average size remained at about 7 feet, 6 inches (2.3 m).

The record length for alligators harvested in Texas is 14 feet, 4 inches (4.4 m), and two alligators reached that size. The first was taken in 1998 in Jackson County by Larry Janke, and the second was harvested in 2003 in Calhoun County by J. V. Dornak. Both were males, since females are not known to get that large. The largest female killed was 11 feet, 3 inches (3.4 m), in 2004 in Jackson County. Female alligators 10 feet (3 m) or longer are rare. I was advised by a former biologist with TPWD that some male alligators are mistakenly recorded as females by hunters who do not make the effort to adequately sex them via cloacal probing.

Selling Hides and Meat

Alligators taken under an alligator hunting license in Texas can be sold either to an individual who possesses a valid wholesale alligator dealer permit or a valid alligator farmer permit. Items made from processed alligator hides such as boots, shoes, belts, wallets, or luggage can be sold by any retailer without a special license or permit (fig. 5.2).

The same is true for packaged alligator meat. However, in order to sell skulls, feet, or teeth, a retail dealer permit is necessary. Although some meat and hides are used commercially in the United States, most end up in the overseas market.

Alligators that are used commercially for meat are skinned and the meat is removed and packaged in processing facilities that also contain freezers to keep the meat until it is sold

Figure 5.2.
Tanned alligator
hides are turned
into a variety of
luxury leather
goods.

(fig. 5.3). Since alligators are not birds or mammals, they are not subject to US Department of Agriculture (USDA) jurisdiction and inspection like beef and poultry are. However, they are inspected by the Food and Drug Administration (FDA).

What does alligator meat taste like? The most common comments that I have heard are that it tastes somewhat like chicken, and I am inclined to agree. It definitely does not have an outright fishy or gamey taste. To me, it does have a slight seafood taste. Glasgow's *Social History of the American Alligator* (1991) goes through quite a discussion of the subject. The first alligator meat that I tasted was at the first

hunt. It was tail meat that had been prepared in a deep fryer as gator nuggets. I do not normally eat deep-fried food, and it tasted much the way that chicken nuggets do—not much flavor in the meat; the flavor was all in the fried coating. Even so, I could detect a difference from an outright chicken taste. In the early 1990s I had the chance to eat alligator meat in many forms courtesy of a dinner prepared by the Florida Alligator Farmers' Association for the participants in a crocodilian conference. I remember the barbecued ribs more than anything else, as they were quite good.

The History of Alligator Poaching in the South

There are two main types of poachers: locals who have a subsistence lifestyle and are trying to feed their families or just get an alligator for their hide or the "heck of it" and those who want to participate in the illegal trade of alligators for profit. At times, these two groups overlap. Before I was educated in the history of alligator poaching, I thought that the two important time periods for poaching occurred when alligators could not be legally hunted because it was prohibited by state law or the Endangered Species Act and after the legal hunt started on a state-by-state basis. After talking to Texas game warden Barry Eversole,

I have regrouped the two eras of poaching into poaching in the high-market era and poaching in the depressed-market era. The important difference of this change in categories is that poaching as a commercial enterprise did not stop altogether with the start of legal alligator hunting. Barry (now a game warden in Fort Bend County, where he was born and raised) told me that during his stint as a game warden in Galveston County alligators were still poached for commercial trade through the late 1980s when the price of hides was high, even though there was a legal hunting season in Texas beginning in 1984.

There were two individuals whom Barry caught in the act of poaching after legal hunt-

Figure 5.3. A commercial processing plant where alligators are skinned and the meat packaged and put into on-site freezers.

ing resumed in Texas. One of them actually met regularly with Barry for coffee. Each one knew what the other was up to, yet they were friends outside the "conflict" arena. Barry also knew that it could be dangerous to deal with this individual if he met him at night in the marsh, especially if he was in the company of other poachers. One night Barry caught him "in the act" and cited him for poaching. When the man died, his son phoned Barry and asked him to attend his father's funeral, saying that his father would want Barry there. He did attend the funeral. The poacher just described, Herman Johnson, has his photograph on the front cover of Melanie Wiggins's book, *They Made Their Own Law: Stories of Bolivar Peninsula* (1990). Barry took a copy of the book to Herman and asked him to autograph the photo. Read his interview in the book. He is a man many of us would have enjoyed meeting for coffee and a chat on a regular basis, just as Barry did. Although Herman spoke about alligator hunting to the author, no mention is made of his escapades in poaching. His colorful family tree was mentioned, however, as he was the great-great-grandson of the cabin boy to the notorious pirate Jean Lafitte. Quite fitting for a modern-day pirate of sorts!

When hides go for $25 to $50 a foot and the market is not being well supplied through legal hunting and farming, illegal hunting takes place. Even though an individual could hunt alligators legally beginning in 1984, it was much more profitable to be able to start hunting them in May and keep on all summer rather than hunt only during the set season in September. Now that the market has remained

depressed long term and is being well supplied with legal hides from the legal hunts and increased farming, poaching is carried out only by individuals sporadically, whose likelihood of getting caught is slim. Overall, alligator poaching is at very low levels at present.

For people with a subsistence lifestyle in the early twentieth century, hunting was necessary for feeding families. These people were used to taking what they needed from the land and were not worried about using up their resources. The problem for alligators began when they became a cash crop. The invention of airboats allowed quick and efficient access into the marsh and a means to overtake alligators. Spotting planes were also used at times. What fueled this interest in alligators? The fashion industry did, beginning in the 1920s. In Florida alone, over 200,000 skins were taken in one year. Obviously, the resource could not keep up with the demand and decreased in number (Reisner 1991).

Louisiana was believed to have lost 90 % of its alligators between 1938 and 1958. Extermination of the alligator was occurring throughout its range. The first state to afford the alligator complete protection was Alabama in 1941. In the 1960s, Arkansas, Florida, Louisiana, Georgia, and Mississippi also gave the alligator protection. Texas was the last state to protect the alligator and did so in 1970.

I have always read and believed that alligator numbers were underestimated at the height of poaching due to their wariness, which explains why alligators were thought to have made a comeback much more quickly than expected. While this premise may be true, it turns out that it is a bit more complicated than that. I also got an education from Barry Eversole about the historical poaching that took place prior to the alligator's rebound. Barry urged me to read the book *Game Wars* by Marc Reisner (1991), which is about the experiences of US Fish and Wildlife Service undercover agent Dave Hall. He said that it would help me understand the history of alligator poaching. The book is divided into four sections, the first of which is a fast-paced and exciting story on alligator poaching. It would make a great movie, and amazingly, it all actually happened.

Various state wildlife agencies and the federal government, for the most part, believed that large-scale poaching was not a problem once the alligator was protected. Many blamed the alligator's continuing demise on habitat loss. Fewer people believed that major poaching was still going on despite no tangible evidence to the authorities supporting that contention. All across the alligator's range, locals were poaching alligators. The locals sold to buyers, and the rise and fall of big-time hide buyers turned out to be associated with alligator populations immediately increasing or decreasing in the years before hunting became legal again.

Poaching was so hard to stop at the local level because of corruption that extended to Fish and Wildlife employees, game wardens, law enforcement, sheriffs, and even judges, and perhaps even higher levels as well. Sometimes people were paid to look the other way, and sometimes kinship was involved: it was not unusual for enforcers of the law and poachers to be related. It is hard to turn in, arrest, or convict a relative. Although the Louisiana

Department of Wildlife and Fisheries and the State of Louisiana played a central role in a lot of what went on in tracing poaching from hunters to buyers, this story is not limited to Louisiana. This fascinating history of poaching and the series of events concerning the major buyers of the illegal alligator hides in the United States really involve the southern "alligator states" as one entity, and Jefferson County, Texas, plays a prominent role.

Leslie Glasgow, a well-known wildlife management professor at Louisiana State University, was hired in the 1960s to head up the Louisiana Department of Wildlife and Fisheries. In the past, the department was known for its corruption, so hiring Glasgow was a very good public relations move. Dave Hall, as a college student at Mississippi State University, used to go to LSU to attend Glasgow's seminars. A student of Glasgow's, Ted Joanen, became one of the best-known individuals in the world for his role in crocodilian research, conservation, and farming. In 1966, Glasgow had no idea that big-time poaching was taking place in his state, which had protected alligators for the past two years.

But in the fall of 1966 he learned from a recently arrested small-time alligator poacher that big dealers were able to operate openly on a large scale, while small dealers were forced out of business (obviously because officials turned the other way in regard to the illegal operations). The Mares brothers and Ralph Sagrera were major dealers for alligator skins in the South but could not be "touched" due to their connections. They were legitimate businessmen and in the top ranks of world fur dealers. The Mares brothers were caught buying the skins through a sting operation involving an individual caught poaching alligators. Warrants were issued for their arrest, and the case was considered to be airtight, but the assistant district attorney did not bring the case to trial when the indictment was called. Months later Glasgow was fired by the same Louisiana governor who had hired him, John McKeithan. (Glasgow's son, Vaughn, describes what he terms the "alligator wars" of the 1960s and 1970s in his book, *A Social History of the American Alligator*. It touches on, and sometimes glosses over, what is talked about in Reisner's book and has additional information about the alligator and its true status at that time.)

Glasgow's firing was probably a blessing in disguise, as he was then appointed assistant secretary of the US Department of the Interior and was now in charge of the USFWS, which had nearly 200 special agents (game wardens at the national level). In 1969, Glasgow transferred agent Dave Hall from Memphis to Louisiana to deal with alligator poaching.

Individuals learned a lesson from the Mares arrest that it did not pay to consort, inform, or work with any government entity that enforced laws on wildlife. The hide buyers became wary and began laundering alligator hides through the last legal place to hunt alligators, Jefferson County, Texas. Hides from all over the South were shipped to Texas and then legalized as Texas alligators. The Mares brothers seemed to have dropped out of the alligator market, and other buyers/exporters moved in and took their place. Law enforcement had no idea who the buyers and exporters were, and it was next to impossible to get any informants.

So law enforcement had to concentrate on catching the alligator poachers.

Because the Endangered Species Conservation Act was passed near the end of 1969, there was a heightened watch placed on all potential alligator hide dealers in the country. The act made it illegal at the federal level to engage in interstate commerce of reptiles and amphibians taken against state law. Another supposedly legitimate large fur and hide business was the Q. C. Plott Raw Fur and Ginseng Company, located in Atlanta, Georgia. The USFWS traced a major shipment of alligator hides in Savannah destined for Japan to the Plott Company. Plott kept records of both his legal and illegal transactions. In addition to the names of over 200 individuals who had sold him alligator hides, he recorded the names of 50 firms in 17 countries that were listed as regular customers. Astoundingly, he had sent out over 127,000 alligator skins (nearing half of what was believed to be left in the wild) in a three-year period.

Immediately after the Plott bust, three people were murdered and two others were left for dead and never recovered. Pappy Wellborn of Beaumont sold Plott many hides from Texas and was suspected of being behind the operation that laundered alligator skins from other states. Two men broke into his home at 2:30 a.m. and beat him and his wife until they were comatose and took only Pappy's business records. Two weeks later another poacher, whose name was in Plott's records, and his friend were found floating in a marsh. Each had been shot numerous times. Then a convicted poacher was found hanged in his cell, and the death did not seem likely to be a suicide. The poacher's best friend, A. J. Caro, was in another prison in Louisiana and feared for his life since he was friends both with Pappy (who was being kept alive on a respirator) and the poacher who was found hanged. Dave Hall had US marshals move A. J. from the parish prison to a federal penitentiary in Atlanta for his own protection.

Dave Hall went undercover as co-owner of a tannery whose owner had been indicted for tanning alligator skins for Plott. A deal was cut, and Dave used the tannery as a means to get illegal alligator hides brought directly to him. When busts were made, he was arrested, went through the courts, and was even put in prison to maintain his cover.

By late 1974, the USFWS thought it had shut down all of the major alligator dealers, as alligators began to rebound in the South, based on annual censuses. These data supported the contention that poaching, not habitat loss, was the main cause of the alligator's decline. Now it seemed that the numbers were going down again, at least in parts of some states. The USFWS learned from Woody Dufrene, implicated with Plott and now out of prison, that Kelly and Klapisch from New York were now running the entire illegal alligator business in the South. Kelly was a buyer of hijacked goods, and Klapisch was one of the world's largest fur traders. They had done business with Plott, as both names were listed in his files.

The operation involved poachers bringing Kelly and Klapisch hides, which were transported to New York, and then most of the hides were shipped overseas. Woody supplied the information that they would be receiving hides at a hotel in New Orleans and the date. Known poachers were seen coming and going from the hotel. The hides were transported from the hotel in rental vehicles, which would not be suspected of harboring illicit cargo. The Feds tailed them with a number of vehicles along the way, and even a spotting plane, to a warehouse in New Jersey. Kelly and Klapisch were both arrested and received jail time that was suspended to probation and fines close to $10,000. Kelly and Klapisch were expected to go back to "business as usual" since sales of alligator hides were so lucrative and the number of USFWS agents was reduced, resulting in the two of them not being watched at all.

Then A. J. Caro came into the picture again in 1976, almost three years after getting out of prison. He knew that at some point Dave would request a payback for having him moved to a different prison and the strong possibility that he had saved A. J.'s life. Ironically, A. J. ended up phoning Dave Hall and begging to be deputized and go undercover, or he was threatening to kill Kelly himself. It turned out that he had been working for Kelly, buying hides and then driving them to New York in a surplus mail truck. Then the hides were loaded on ships headed for Japan. A. J. had brought his girlfriend along with him on the last trip, and he suspected she had been drugged and raped by Kelly when she was left alone with him.

It was decided that A. J. would transport more hides to Kelly and bring another "girlfriend" along this time who was really an undercover USFWS agent named Marie Palladini. Dave Hall thought that if A. J. had another woman accompany him, it would indicate to Kelly that the last girlfriend had not remembered anything. An interesting aspect of A. J.'s preparation of going undercover is that Dave Hall had to personally pay for the hides that A. J. bought for the sting operation by loaning him his personal truck, credit cards, and spending money. Obviously, this whole operation was more than just "doing his job" for Dave. After A. J. and Marie arrived in New York, the van with the hides was temporarily parked inside the garage at Kelly's mother's house. Dave Hall needed to know who the buyers of the skins were going to be. A. J. found Kelly's records under the right front seat of his car and ripped out the pages that he needed. The hides were to be shipped to a company in Tokyo, but how did the sale to the Japanese work?

After leaving Kelly's possession, the hides were sent to a freight forwarder, who took them to the ship at the last minute to avoid a lengthy customs inspection. Once the ship was beyond the two-hundred-mile limit, Kelly and then A. J. would be paid. In this manner the Japanese minimized being linked to Kelly and Klapisch, who were being watched by the Feds.

A. J. ended up working undercover for the DEA (Drug Enforcement Administration) in the poaching operation and had to stay behind when things got dangerous to make their case. He ended up phoning Dave Hall from Kennedy Airport, while he was hiding behind a Delta Airlines counter, fearing for his life. Dave Hall wired him a plane ticket, and he actually lived with Dave and his family for a month. Kelly was arrested and convicted by A. J. and

Marie's testimony. He turned state's evidence and brought down a number of other criminals unrelated to the alligator case, including a counterfeiter, brothel owner, and some people in the illegal drug trade. Kelly was caught again trading in alligator hides a couple more times and then started lying low. And we know that finally the alligator rebounded throughout its range and that a legal hunting season recommenced, state by state. Keeping track of what everyone was doing to keep legal seasons going on indefinitely was Dave Hall's goal. His comment was (in reference to potential poachers), "So I go through here now and then to let em know they're still being watched and to learn who's killing too many. Now that they're learning the value of *management* . . . they're informin on each other all the time. Self-policing, I guess you could call it" (Reisner 1991, 71).

Alligator Farming

I'm an old gator hunter from way back. It's better than cattle. You don't have to feed 'em, vaccinate 'em, or worry about any kind of a fence, or nothin.' So, I mean, it's just profit, you know?

—Herman Johnson, quoted in Melanie Wiggins, *They Made Their Own Law*

Alligator farming is a larger commitment than alligator hunting, something like marriage versus a single date. Basically one can go out for a weekend and harvest an alligator after having taken care of the appropriate licensing. However,

Nutritionally Complete Diets for Crocodilians in Captivity

I see problems with breeding crocodilians in farming situations in part by how the animals are configured in regard to having home ranges and adequate areas to thermoregulate properly when being kept with other adults. However, I am interested in nutrition and think that feeding pelleted foods with some frozen food (such as fish) is not an adequate diet long term and is the biggest problem in the lack of breeding success. It is one thing to raise young alligators for several years on such a diet to bring them to harvest size. It is entirely different to keep breeding animals for many years and have them successfully breed on a regular basis. Bill Ziegler is formerly of the Miami Metro-Zoo and now senior vice president of Animal Collections and Care at the Brookfield Zoo. He believes that his success with breeding crocodilians in captivity is due to diet. His commentary here is something that I consider to be of use to alligator farmers.

"I view diets in a simple manner. The more whole food and bigger the variety, the better off you are in meeting the animals' needs. If you have a safe supply of crayfish, apple snails, and other crustaceans individually, or crushed and added to other items for bigger crocs are also good, just be sure your source has tested them for parasites. The addition of supplements was a matter of observing my animals, noting egg density, when they laid, fat deposit buildup around the neck and arms, color of teeth, and so on. Nice whitish opaque teeth means a good diet. I can still go see animals today that had discolored yellowish teeth or hardly any teeth at all in a lot of institutions. In the wild you'll never see that in a healthy animal.

"You can also overfeed or feed the wrong things despite your best intent. I tried to stay away from frozen food as much as possible and making fish no more than 5 to 10 % of the total diet. Most fish came to us frozen, which meant I used it even less. Also, add greens to the diet; you can do this by stuffing them down rats and rabbits. In the wild these guys eat whatever they grab, including the plants, et cetera, that may stick to the prey, to say nothing of the veggies that are already in the gut of the prey to begin with. If you feed them just fish, they get none of that, and if you feed them lab animals, the lab animals were fed a processed diet designed for its absorption, so little is left in the stomach. They need the roughage from a fresh meal of greens the prey just ate. Kale, collard greens, little amounts of spinach, and romaine lettuce are good. I know many people who will pooh pooh this info, but it worked for me in consistently breeding."

farming alligators requires a minimum of providing a facility in which the alligators are housed, fed, and cleaned up after and then harvested and sold. Depending on the facility, breeding may take place on-site, eggs may be incubated and hatched, and the young grown and harvested. There is also the power bill for heat and lights. Feed can be chicken, wild nutria, a processed feed, or a combination. So, alligator farming takes considerable time and money.

Mark Porter is at the pinnacle of the farming and overall alligator processing operations in Texas. Although he makes more money from alligators commercially than anyone else in Texas, it is just a part-time job. His full-time job is in Baytown at the Chevron chemical plant as a production operator. His wife, Kay, is a schoolteacher. I know from personal experience that being a schoolteacher can involve staying up late at night preparing lectures, grading tests, and so on—far beyond the school day. She and their daughters have also helped with the family alligator business. They do have employees who help them clean their house, work the alligator farm, and maintain their yard and grounds. They work continuously and have very little time to take off. Having employees just helps them get things done; it is not a way for them to get by not doing work themselves and just doling out paychecks. It is survival and a harder way of life than most people would choose in terms of the workload. Most people would not choose to work so hard to make money.

Recent History of Alligator Farms in Texas

The first licensed alligator farm in Texas began in 1986 with 206 alligators. None of the three farmers who started the first farm are still in business. Twenty licensed farmers with a total of 1,414 alligators existed by 1988. This year was also the first for farm-raised alligators to be harvested. A total of 20 animals ranging from just over 4 feet to just under 5 feet (> 1.2 to < 1.5 m) in length were harvested by a single farmer. By 2002 there were 26 farmers with over 39,000 alligators in Texas, and about 15,520 farmed animals were harvested (approximately 5,000 fewer than in the previous year).

Obtaining Breeding Stock

If a farmer wants to try to breed alligators on the farm rather than get eggs from wild nests or buy hatchlings, then adult alligators are necessary. Since one male can breed a number of females, most adults need to be females in order to maximize the possible number of nests. Adults can be purchased, or some of the juveniles on the farm can be grown to adulthood. Al Janik has a unique way to get breeding stock. As a nuisance alligator hunter and a farmer, he can use any nuisance females as part of his farming operation.

There is a potential for using artificial insemination to increase successful breeding on farms. In the 1970s and 1980s semen collection was tried, using electroejaculation on crocodilians (and even Galapagos tortoises at the Houston Zoo). The animals undergoing this procedure were not having a good time. They had to be restrained and underwent a lot of stress. This method also failed to produce the significant amount of sperm required for reproduction. Jesse Krebs of the Omaha Zoo noted that "one million sperm are required to have successful fertilization with artificial insemination in crocodilians." He went on to say that "the only way to get one million sperm is to kill the animal, remove the vas deferens, and strip the sperm from there." Jesse and his crew in the Reptile and Reproduction Departments at Omaha Zoo have developed a method to collect semen using training and a collection technique that requires no restraint and is not invasive. As Jesse puts it, "Participation by the alligator is strictly voluntary." One sample collected by zoo staff produced a concentration of 1.5 million sperm with 32.6% motility (that's considered good), and a second sample produced 1 million sperm. Each sample was taken in March 2007 from a different male, both approximately 9 feet (2.7 m) in total length.

Egg Collection from Wild Nests

Overall, breeding in the wild works out better than a captive-breeding program. Historically, alligator farms of the past and present have been supplemented by having eggs and/or hatchlings from the wild to keep them going.

HOW EGG COLLECTION WORKS

A person who wants to collect eggs pays the state of Texas $50 per nest stamp and the landowner $200 for each nest. The process starts by flying over an area in a helicopter to look for nests. When a nest is located, a PVC pipe is used to mark the spot. Then the egg collector will go back, usually by boat, and retrieve the eggs (fig. 5.4a). Two people are involved in the process, one to collect the eggs and the other to physically ward off the female if she shows up and becomes aggressive while the nest is robbed of eggs. Each egg is marked according to its position/orientation within the nest so

that it can remain in the same orientation to maximize survival of the embryo. Flooding and predation can result in a loss of eggs from wild nests before the egg collector can reach them. Mark Porter lost 20 nests in one year due to flooding at a loss of $20 per hatchling.

HISTORY OF EGG COLLECTION IN TEXAS

The egg collection program began in 1988 from nests in Chambers, Jefferson, and Orange Counties. In 1991 and 1992, it was expanded to Brazoria and Liberty Counties, respectively. In 1994, Calhoun, Colorado, Fort Bend, Galveston, Henderson, Jackson, Matagorda, Victoria, and Wharton Counties were included in the program.

James Broussard was the contractor who conducted the 1988 first harvest of alligator eggs and subsequently incubated them. He took a total of 2,261 eggs from 65 wild alligator nests in Jefferson County. The highest number of eggs in one nest was 52, and all of the eggs were fertile. The estimate on the embryo age of the eggs at the time that they were taken ranged from 10 to 40 days. Of those eggs, 2,146 were set up in incubators and 1,872 hatchlings resulted. Approximately 87% of those eggs produced healthy hatchlings. Overall at Texas farms, hatching rate was first at about an 80% success rate. Due to better culturing and facilities, the rate of hatching success is now at about 92%.

Hatching the Egg

Incubators play an important role in how many eggs that are set actually hatch with viable offspring. As it is for an incubator for any type of eggs, maintaining a desirable temperature and

Figure 5.4.
(a) Herbert
Oreschnigg
collects eggs
from a wild nest.
(b) The eggs are
then taken to
his incubator,
which is a room
designed to
have a constant
temperature.
The young
are sold to an
alligator farm
after hatching.

a

humidity is the goal. James Broussard's incubator that first year of egg collection cost about $6,000 to build. The temperature and humidity were set by heating water in the incubator to the desired temperature of 88°F (31°C) and then insulating the incubator with about 4.5 inches (11.4 cm) of foam insulation. The humidity level stayed at just above 90%. Just in case the temperature could not be maintained within limits of 86°F to 90°F (30°C to 32°C), a high/low alarm system was utilized as a backup. If the high-temperature alarm went off, a cooling fan came on. If the low-temperature alarm sounded, a heater turned on. The incubator never went outside the critical limits that first year, so this simple and relatively inexpensive incubator did its job very nicely.

Prior to being put in the incubator, each clutch of eggs was placed in one or two hardware cloth baskets (depending on the number of eggs). The eggs were surrounded by nesting material to simulate conditions inside the nest, including providing the source of the bacteria that has been found to degrade the eggshell and facilitate easier hatching.

With the proper conditions in an incubator, survival in captivity is increased over that in a wild nest by protecting them from large predators such as raccoons or hogs and fire ants, by culling nonviable eggs throughout incubation, and from aiding in the hatching of weaker individuals from eggs or monitoring/protecting eggs that are not ready to hatch with the rest of their clutch. After hatching, the

b

young alligators remain in the incubator for up to a day (fig. 5.4b). Those that have not fully absorbed the egg yolk or are weaker may be kept several days in the incubator. In this case some of the nesting material may be removed to allow them to move more freely within the confines of the basket, whose top is secured if the sides are short enough that the hatchlings may crawl out. Those individuals with incubators may sell some of their hatchlings to other farmers who are not set up to incubate eggs.

"Growing Up" the Hatchlings: Recipes for Success

Hatchlings may be raised in indoor or outdoor facilities. Both provide the alligators with habitat including land (some type of substrate free of water) and water. There are advantages and disadvantages to both. Indoor alligator farms may raise young alligators with reduced lighting (i.e., under conditions of semidarkness) in order to reduce fighting among each other, which results in better-quality hides with less scarring. Temperature-controlled rooms result in the alligators eating year-round, which in turn increases growth rates. Normally an alligator grows about a foot its first year of life, but can reach to 3 to 4 feet (0.9 to 1.2 m) in the same time frame if kept in warm temperatures and fed continuously (they normally go off feed in the wild and in captivity with a change to lower temperatures). This spectacular growth rate is more likely attainable with males, whose growth rate is faster than that of females. Predation from natural predators is also eliminated. It costs a lot to go this route but is usu-ally well worth it if the farmer intends to put the time and effort into running an operation of this type.

Hatchlings raised in outdoor conditions are exposed to the sun and vitamin D. If given the correct water-land ratio and they are not over-crowded, they can do very well. Raising them outdoors also requires safeguards that keep predators out (both from above and digging in). Heated ponds can be utilized if desired.

The outdoor enclosure used by James Broussard was about 30×50 feet (9.1×15.2 m). A fence was made from corrugated metal roofing arranged longitudinally around the perimeter and sunk in the ground about 6 inches (15.2 cm) deep to secure it. Electric fencing and netting surrounded the enclosure to deter predators. The land-to-water ratio within the pen was about equal, and fresh water was circulated through the tank during the day (Broussard 1988).

Both indoor and outdoor facilities involve a larger number of animals in closer quarters than they would be in the wild, and this can cause problems in terms of overall cleanliness and avoidance of disease. Pathogenic bacteria and fungi, which may be present in low levels on alligators in the wild, can be a major problem in captivity, where they can build up in higher levels within an enclosure. Indoor facilities are usually easier to disinfect than an outdoor enclosure. However, having an outdoor pool that can be easily drained and disinfected makes a big difference. Being able to pull out on dry land and bask in the sun helps in keeping the animals healthy.

Farming facilities are subjected to regular checks by TPWD, which includes such factors as cleanliness of the enclosures, overall health of the alligators, and adherence to stocking quotas within a certain size pen. Mark Porter notes that he changes the water in the indoor enclosures once or twice a week, depending on how dirty it gets. He tends to stock below the level allowed by TPWD. It pays off in terms of healthy animals that grow quickly. Mark's grow-out building is set up into 18 different compartments, which he calls ponds. The ponds are water filled with dry areas on the sides and vary in size. As a group of alligators grows, it gets moved into a larger pond.

Age of Harvest

The more time that an alligator is kept in captivity and "grown up," the more money is expended to raise it. Louisiana controls the market prices that Texas is paid, as hides go to Louisiana. At the 3-foot (0.9 m) size, there is a higher profit margin than the 4-foot (1.2 m) size due to less overhead. Louisiana alligators generally get harvested at the 3-foot size class. Texas alligators get harvested at the 4-foot size class, as the buyers in Louisiana refuse to buy the Texas alligators at the 3-foot length. The Texans are literally "shut out" of the 3-foot market by loyal buyers, who are looking out for the market in their state. What is so ironic about all of this? Remember that in the wild in Texas, hunters cannot hunt alligators less than 4 feet in length and usually do not want them that small anyway. Small leather goods are made from these captive-bred hides, which have a smaller pattern than the adults and are more perfect than wild hides. Farmed alligator meat goes for a higher price than meat from the larger wild alligators. So in the alligator farming business, smaller is better.

Types of Farms

Mark Porter's alligator business involves everything but exhibiting alligators (Gator Country is the only farm to do that). He collects eggs, incubates them, raises the hatchlings, and buys additional hatchlings to raise. He also has a large pond in back with adult alligators that may nest and produce additional eggs. He has a separate building where he processes harvested alligators and where there are freezers to hold the meat (Franklin 1998).

The alligator's comeback has been a means to generate income that did not exist (legally) since the reptiles were overhunted. While not an important part of the total income generated in Texas (especially in direct comparison to the oil business), it makes a positive contribution to the livelihood of individuals on the Texas Gulf Coast in both the coastal and inland areas where large populations of the alligators exist.

6 Alligator-Human Interactions

In my boyhood days before these reptiles had been disturbed by hide-hunters I came in contact with them constantly, and seeing them was such an every-day occurrence that no unusual notice was taken of them by the children playing and swimming in the streams. They were looked upon as part of our natural surroundings, and we paid no more attention to them than we did to the flocks of birds about the place.

Among the earliest remembrance of my childhood is running down with my brothers and cousins and other small boys in the warm summer afternoons to the boathouse to swim; each boy trying to see who could get in the water first. The bayou is about one hundred and fifty feet wide at this point and about ten feet in depth, with several shallow streams coming into it above and below the place where we swam. It is a tide-water stream and most of the time quite salty. In these days alligators in the streams about the place were more than numerous, and of course, boy-like we always took great pleasure and not a little excitement in seeing how many gators' we could call around us during our swim. We would attract them by imitating the barks and cries of dogs and by making loud popping noises with our lips, as these sounds seemed to arouse the gators' curiosity, and they would come swimming to us from all directions. We had no fear of them and would swim around the big fellows, dive under them and sometimes treat them with great disrespect by bringing handfuls of mud from the bottom and "chunking" it in their eyes. Sometimes when the tide was low we would surround on three sides a big one that might be lying on the edge of a flat, and create such a commotion splashing and jumping in the water that the alligator would crawl out on the mudflat, and we would follow him "chunking" great handfuls of soft mud in his eyes and open mouth, and on several occasions in this manner we actually overpowered them, and after tying their jaws, dragged them to the house.

—E. A. McIlhenny, *The Alligator's Life History*

Alligators and human beings are bumping into each other more frequently in Texas nowadays. Alligators have made a comeback since they have been afforded protection, and their numbers are continually increasing. At the same time, the amount of habitat available to them and the quality of that habitat are continually dwindling as a result of the encroachment by an increasing human population. We are faced with a predicament in which humans and alligators are often vying for the same land. Humans go to areas inhabited by alligators for recreation (hiking, biking, fishing, hunting, boating, and swimming) or to live. Alligators migrate to areas inhabited by humans as they actively move through to get somewhere else, as they make rest stops along their journey, or as they set up residence.

I saved a newspaper clipping from the *Houston Post* dated July 15, 1988: "Coincidence or Not, Houston Host to Gators Galore." The article noted that Houston's Animal Regulation and Care division had "picked up more alligators in the past six weeks than in the past five years." One of the four alligators mentioned in the article was an animal 5 feet (1.5 m) long that was sighted in White Oak Bayou and reported by joggers who feared for their dog's safety. If my interpretation of the article was correct, this animal and the others described were relocated to the J. D. Murphree WMA in Port Arthur. This scenario is very different from the occasion when two subadult alligators turned up in Houston bayous several years later. Although these alligators hit the local television stations' 10:00 news, they were allowed to remain where they were because

bayous are their natural habitat and they were not considered to be a threat to humans. Now that alligators are becoming more common, TPWD has stopped relocating every alligator that it gets a phone call to remove. There is the realization that alligators and humans are going to have to learn to coexist.

The general public, for the most part, has not reached the stage of comprehending that unless an alligator is large, aggressive (usually habituated to humans), and in close proximity to human dwellings, it will be allowed to remain where it is and not be disturbed. Furthermore, since it is a protected species, it is illegal to catch a baby alligator to keep as a pet, harass an alligator in any way, or kill an alligator unless an individual is signed up for a specific alligator hunt. I have spoken with fishermen whose favorite spots where they had gone wade fishing for years are now home to some large alligators. They were in a state of disbelief that their phone calls to TPWD did not result in the alligators being immediately removed. The idea that they should just stay out of the water did not sit well with them, because they thought that their rights were being violated. Homeowners who have a pond in the country and fear for their dog's safety when an alligator shows up find it a hard pill to swallow when a game warden refuses to remove an alligator that is not deemed dangerous, while a county animal control officer tickets the owner for having a dog "at large." The landowner is furious that wildlife has a right to come and go on his land but that the dog cannot legally run loose.

Alligator Problems in Human
Habitations and Parks

▼▼▼▼▼

Clashes between alligators and people occur
primarily when alligators lose their fear of
people. There is a big difference in the behavior of wild alligators and those that are habituated to human beings. If you stand on a shoreline and view a wild alligator from a distance,
it will typically "just sit there" or give a quick
splash and "dive under" the water if it becomes
uneasy from being scrutinized. If you look at
a habituated alligator in the same situation, it
comes toward you, possibly even aggressively.
Big difference! Humans make problems with
alligators for the most part. (However, there are
some specific problems that occur associated
with nesting, discussed later.)

Feeding Alligators
The most common and fastest means for alligators to lose their fear of human beings is
for people to feed them, whether with "people food," a freshly caught fish, or roadkill.
Humans can unintentionally feed alligators
as well. This often involves an alligator being
attracted to a fish that is hooked on the end of
a line or fish on stringers. Another common
means to "accidentally" feed an alligator is to
clean fish and throw the heads and other discarded parts into the water. Some alligators are
too shy to make the move to grab a fish off a
line, but those that do generally begin to make
a habit of it and then become bold enough to
come onshore and grab stringers (or even try
to grab them from a boat). Keep in mind that
"top-water" fishing lures may attract alligators,

just as they attract fish, because they mimic
natural prey.

Dogs
Dogs and alligators do not mix. Waterfowl
hunters have always used caution when sending their dogs in the water to retrieve birds
where alligators range. Dogs are prey size and
are considered to be prey by both wild and
habituated alligators. Our domestic canines
are no match for alligators and are easily
taken down when standing in shallow water
or swimming. My observations are that dogs'
behavior is a large part of the problem. I pro-
pose that dogs elicit a territorial response from
alligators when they do not necessarily elicit a
prey response. Alligators chase other alligators
out of their territory, and my observations and
the literature suggest that dogs may sometimes
be treated as invaders, not prey. There is a
documented case in Florida of an alligator that
lived at a golf course that did not eat, but killed,
a retriever-type dog when it entered the water.

Dogs often create a lot of noise by splashing in the water, and a typical response of a
dog to an alligator would be to bark. Based on
the social behavior of alligators, this might be
the equivalent of a bellowing display, which
can be perceived as a display of dominance by
other alligators. My neighbors have a golden
retriever that likes to go wading and swimming. They observed it being chased from a
pond by an alligator and found bite marks on
its foot. I have a 100-pound bouvier des Flan-
dres that has had some exposure to alligators.
His behavior (representative of the breed) is
typically that of a big, stubborn draft horse. He

is not a barker and is usually quiet and deliberate about his movements. He does not seem to elicit the same negative response from alligators that the neighbor's dog does. He also would be unlikely to elicit a prey response, unless he was in deeper water wading or swimming and/or the alligator was extremely large. Would I bet my dog's life that it would not be attacked by any alligator? No! But I suspect that most dog breeds (and individual dogs), with a high degree of predictability of which dogs, would elicit an aggressive response from an alligator.

Dogs that are strictly pets often accompany their owners to parks and nature trails where dogs are allowed. It is imperative to follow safety regulations. Even when signs are posted about alligators being present and/or dogs being kept on leashes, tragic events occur when the owner disregards safety precautions. The dog is either allowed to run loose and gets into the water while the owner yells at it to get out, or the owner feels sorry for the dog being out in the heat and allows it into the water while it is on leash. Too often a large alligator takes advantage of the easy prey and takes the dog underwater, and it never surfaces again, at least not alive. Shortly after Brazos Bend State Park opened, a poodle and a Weimaraner were killed by alligators when their owners let them swim in a lake there. As Dennis Jones, who worked for numerous years at Brazos Bend State Park, says, "A wet dog is a dead dog!"

Sometimes people try to follow the rules to the letter but leave their common sense behind. I observed a woman fishing at Brazos Bend State Park some years ago on the north side of Elm Lake. The shoreline going into the water was very shallow, and there were several large alligators in the vicinity. At that particular time, there was a large alligator known to come onshore and grab stringers of fish. The woman had tied her dog to a tree near the shoreline. I told the woman that she needed to untie her dog and get it away from the water, as it was going to be alligator bait. She protested that she was following the rules by keeping the dog on a leash. I tried to explain to her that tying the dog up was worse than letting it run loose, because it would have no chance to run away from an alligator. I am not sure if I ever succeeded in getting my point across to her—she was mired in the rules!

Amos Cooper, biologist with the TPWD Alligator Program, related an experience with a woman who moved into a new house to be closer to nature. She had a small dog and was concerned about it being eaten by an alligator. She wanted any alligators near her residence removed. She did not want to put up a fence to contain her dog, as Amos had suggested. He then asked her what she would do if a pit bull moved next door. She replied that she would put up a fence!

Humans at the Water's Edge

Large alligators can perceive young children as prey when they are in shallow water or at the shoreline. Children bending over to closely examine something at the water's edge are not paying attention to what is going on (i.e., an alligator watching them), and their prone position makes them look smaller and easier to attack. Even older children or adults are more susceptible to an attack by an alligator, or having an alligator come onshore to steal fish from their line or stringer, if they are sitting or bent over. The best thing to do if you think that an alligator is sizing you up and starting to come toward you is to make yourself look as tall as possible and calmly back away from the water. If you have a child or dog that is attracting an alligator's attention, pick the child or dog up, and if possible, place the child on your shoulders. This makes you look even bigger, and the child less accessible. This strategy works with a number of wild animals. A human in an upright position is very intimidating!

An example that highlights the concept of size involved Fort Bend County game warden Barry Eversole and his three sons, who were fishing in the presence of alligators. Barry noted that when he or his two oldest sons passed along the shoreline where they were fishing, an adult alligator was in sight and did not have any discernible reaction to their presence. However, when his youngest son (then four years old) passed by the alligator, the alligator exhibited considerable interest. Barry tested the premise that it was the size of his youngest son that was eliciting the response. He found that through repeated testing by having himself and each of his sons pass by, that only his youngest son caused a response from the alligator, prompting him to believe that the relative small size of his youngest son was directly responsible for the interest of the alligator.

Another story about a visitor at Brazos Bend State Park defies common sense and illustrates what not to do around an alligator. A woman was spotted emerging from one of the lakes in full scuba gear. When questioned by the assistant park superintendent why she would get into the water with alligators, wearing a costume that prevented alligators from identifying what species she was, and swim in a prone position, she seemed to take the whole episode very lightheartedly. Her response was that she was a professional diver (as if that would make a difference to an alligator). Obviously, if being in a "bent-over" position is not good, being totally prone is no better, especially in disguise!

Boating

Boats with a noisy engine, such as motorboats and airboats, tend to scare off alligators. Nonmotorized boats, such as canoes and dinghies, may actually attract them. I have heard stories from several people, who are used to alligators and live in southeastern Texas coastal areas, concerning alligators attacking or biting their nonmotorized boat. I have spent a lot of time in such boats in alligator territory and did have an experience in which a large alligator made an aggressive move toward the boat and bellowed. The low profile of the boat, coupled with its occupants being seated and no noisy motor being present, probably elicited a territorial reaction from the alligator. I would consider this a rare occurrence, maybe one out of 300 trips that I made into "alligator-infested" waters that resulted in any kind of adverse response from an alligator. The individuals who related similar experiences to me, including their boats being bitten, were all frequently in water bodies where alligators occurred, so these were unusual experiences for them as well.

Wading

I have waded without problems in waters where alligators are found. This does not mean that I would do something as foolish as walk through the water up to the nest of an alligator that is aggressively guarding it and not expect to encounter trouble. It simply means that under most circumstances alligators do not attack humans and that someone who is experienced with what "evils can befall them" in a marsh can safely work in the water to conduct water sampling, fix a water control structure, or wade in the marsh to collect botanical specimens without a problem from alligators. In the late 1980s when I was ready to begin a mark-recapture of juvenile alligators at the J. D. Murphree WMA, we attempted to see if I could conduct captures during the day. I went with TPWD personnel, who were working on a water control structure, and we stood in waist-high water for some time. Some adult alligators approached us out of curiosity, without any type of aggression. We attempted to catch some juvenile alligators that came close but managed to capture only a couple of them, so we realized that night captures would be much more profitable in terms of numbers.

A note of caution: It is always best never to be alone in the marsh when having to get into the water. Know where alligator holes and dens are, and never wade near them. Waist-high rubber waders can be dangerous if one hits a deep spot, as they can fill with water. I once was wading across from one island to another and hit a deep spot when wearing knee-high boots. I feel reasonably sure that I hit an alligator hole and that the den was at the end of the island.

I went from water that was an inch below my boot tops to waist-deep water in an instant. It was difficult to maneuver in the water with my boots weighing me down. The situation could have been dangerous if I had been wearing waders that filled with water or if I had encountered an alligator.

Basking Alligators

People tend to look at basking or sleeping alligators on land as lazy, as unreal, almost as stuffed animals (fig. 6.1). Dennis Jones was discussing with a woman an alligator basking on a hiking trail at Brazos Bend State Park. She did not seem able to perceive what was really going on with this animal and how she needed to act around it. Dennis was unable to convince her that it was just "taking it easy" and that she needed to respect and not disturb it. She somehow was not convinced, and right in front of him, she grabbed the end of the alligator's tail. The alligator was very startled and reacted. Luckily the alligator did not bite the equally startled woman, she realized how foolish she had been, and Dennis was able to get control of the situation.

A person may even walk right up to an alligator without realizing it if the animal is hidden by tall grass or emergent aquatic vegetation (fig. 6.2). The fact that it is possible to walk right up to a basking or sleeping alligator is a potential problem no matter what the circumstances are. It makes no difference if the alligator is wild or habituated to humans. It is being caught off-guard, and its space is being violated. Therefore, it can react unpredictably.

Figure 6.1. A potential problem is created when people who are inexperienced with alligators walk right up to one of these large reptiles.

Flooding and Drought

A drastic change in water levels can place humans and alligators together in the same location, which would not occur under normal circumstances. Flooding can displace alligators, other wildlife, and even humans from their homes. A number of years ago when Tropical Storm Allison came through the Houston area, a large alligator was filmed in the same rushing floodwaters where rescuers were trying to remove people from their rooftops by boat. Drought provides situations where alligators may move from an area without water to an aquatic area associated with humans or where people may be able to walk on land that was

formerly underwater right up to an alligator hole with a den. The amount of water present in the hole may be insufficient to allow the alligator a means to feel that it can hide or get away, especially if the den is visible. Sometimes even the alligator hole dries up, and the alligator is confined to the den. Some alligators may be very passive under these conditions (fig. 6.3a), while others aggressively keep intruders away with lunges and open mouth (fig. 6.3b).

Nests and Pods of Offspring

Nests present a special problem. Unless an individual knows what an alligator nest looks like, it is possible to go right up to a nest with-

Figure 6.2. If alligators on land are obscured by vegetation, a human can accidentally get too close to the alligator without realizing it.

out realizing it. Sometimes nests can be a considerable distance from the water, thus are unexpected. They can also occur in a dimly lit wooded area where it is difficult to see a nest or an alligator guarding it until one's eyes adjust to the light. I have had a couple of experiences when I was walking through heavily wooded areas looking for alligator nests after having been in bright sunlight. In one instance, my eyes adjusted to the light just in time to see an alligator sitting on top of her nest. In the next moment she was flying toward me with her mouth wide open, off the nest, as I rather hurriedly backed away.

In the second instance, a student intern and I waded from one island to an adjacent one where we suspected that a very aggressive alligator had a nest. The intern watched the alligator from the shoreline to alert me if she started to swim to the island. I proceeded to go through the dense brush and suddenly saw the nest. I yelled to the intern that I had found the nest and quickly yelled again that the female was coming at me (somehow she had sneaked around the other end of the island without being seen). I pushed myself through the dense brush, tripped on a tree root, and fell flat on my face! I grabbed my snake hook, which had fallen beside me, and pushed myself up with all of my strength and ran, without looking behind me. I made it to the backside of the island out in the open and immediately spied a pod of

Figure 6.3. (a) Alligators can be approached at den sites during drought. Some are quite passive under these circumstances. (Photo by Dennis Jones) (b) Other alligators are very aggressive. (Photo by Ed Farrington)

a

b

yearling alligators that we had seen earlier. I realized that these belonged to the female and, as I ran a safe distance away from them in the realization that their mother would soon be there to protect them, called to the intern and conveyed to her the additional danger of the young alligators. Somehow she made it to the spot where I was waiting without meeting up with the alligator. This is probably the most dangerous encounter that I have had when doing field research with alligators. I always try to have a plan for an escape route. In this case, we were outsmarted by the alligator and ended up being trapped. This experience also shows how "forgiving" alligators are in comparison to most other crocodilian species. If the female had been a Cuban or saltwater crocodile, I would not be here to tell the tale.

There is a variation in how alligators react when their nests are approached that seems to be highly correlated to how familiar alligators are with humans. I want to make a distinction between "familiar with" and "habituated to" because I am not convinced that all alligators that actively defend their nests are habituated. At Brazos Bend most nesting females would aggressively defend their nests, yet not all of them behave aggressively at other times of the year or show other behaviors of habituated or nuisance alligators.

When I first began to examine nests at Brazos Bend in 1987, none of the females were aggressive toward me when I approached, and even opened, the nests. Most of them gave a quick splash in the water and submerged. A few stayed in the shallow water near the nest with just their heads showing but made no attempt to interfere with my examination of the nests. In 1990, two females nesting within a few meters of each other, at a lake that was frequented by fishermen, exhibited aggressive behavior when their nests were approached. It involved coming out of the water toward the nest and, then if the intruders did not leave the vicinity, opening the jaws and lunging (fig. 6.4). One of the females climbed on top of her nest and lunged forward from there. The females would tend to run after any intruders for a short distance after they retreated.

In subsequent years, an increasing number of females exhibited the same behaviors in defending their nests until only those alligators in the remote areas of the park did not try to chase us away. The park opened to the public in 1984, and the number of visitors increased to about 100,000 per year. The alligators got used to seeing human beings and the annual "nest checks." I did observe that the areas with the highest numbers of park visitors were the first to have their resident alligators turn aggressive at the nests. However, I believe that my repeated visits to the nests several times each season, and subsequent seasons when the females nested again, contributed to their behavior.

At the Murphree WMA I examined a number of nests for the TSD study in 1987 and 1988 that were accessible only by airboat. In spite of the engine noise, and the fact that

Figure 6.4. Female alligators may protect their nests from human intruders by using open-mouthed lunges.

the alligators probably had had very limited access to humans in their past, one alligator did exhibit aggressive behavior toward us when we were in the nest vicinity. It is unknown if this particular animal might have had prior experience with humans, if it was her individual temperament, or if it had something to do with her hormone levels.

Young alligators giving the "umph-umph" distress call often have their mother, or even another adult alligator, respond to their cry. I once saw a young alligator in the water near the shore and did not see an adult nearby. I was curious if there was an adult in attendance, so I stood about 3 feet (0.9 m) away from the water near the juvenile and gave the "umph-umph" call. I figured that if there was an adult keeping watch, I would probably get its attention, but I was just far enough from the water to easily back away if an alligator emerged from the water. One did, and it raised its head and front part of its body out of the water, then slowly lowered itself back in the water as I backed away. It then stayed in the water with just its head showing but turned its head as I walked near the shore, watching my every move. The moral of the story is to make a wise assumption that every juvenile alligator has a guardian adult.

Alligators That Are "Trapped" in Human Habitation or Injured

When an alligator ends up in a swimming pool, sewage treatment pond, or other water body that is in close association with humans and is not suitable habitat, the animal is often very stressed, dehydrated, and somewhat thin and beat up and feels trapped when surrounded by humans. An animal in this situation that is unable to escape or hide will often hiss, open its mouth threateningly, and lunge. If it is injured, this tends to intensify its aggressive behavior.

Alligators Injured by Human-Made Structures

The most common way for alligators and other wildlife to get injured or killed in association with human-made structures is to get hit by cars when crossing roads (fig. 6.5).

Other sources of injury or death are from monofilament fishing lines and getting stuck inside a small opening such as a drainage pipe. When a fisherman cuts a line after an alligator grabs it, the hook and a length of monofilament line remain in the mouth. This does not kill the alligator, but it does have to live with a hook in its mouth, and there is undoubtedly some trauma and discomfort at the site until it heals over. There is also the potential for the monofilament line to get caught up on something to such an extent that the alligator cannot break free or, if it does, again suffers an injury. An alligator at Brazos Bend State Park at the 40-Acre Lake fishing pier repeatedly went after fishing lines and had quite a collection of hooks in its mouth. At the same pier an alligator got tangled up in a high-strength monofilament line meant for catching big saltwater fish, such as swordfish and sharks, and drowned. On another occasion, an alligator drowned at this pier because it tried to climb through the railing into the water and got its hind legs caught in a small space between the rails while its

Figure 6.5.
Alligators may
be killed while
crossing roads.
Alligators that are
migrating or have
been relocated
and are trying to
return home are
especially at risk.

head was still in the water (fig. 6.6). It drowned because it could not hold its head out of the water and breathe.

Amos Cooper of the TPWD Alligator Program cautioned that gated water control structures with steep sides pose a drowning hazard to alligators. The one that he is familiar with had alligators gain access to it when the gates were open or they climbed over the gates. Then they were unable to climb out since they were in deep water and were forced to swim or float continuously to keep from drowning. Several alligators drowned because they were not discovered in time. Others had to be hoisted out with cranes.

The March 2005 issue of *Herpetological Review* shows a graphic photograph of an alligator in Florida that died when it got stuck while trying to walk between rocks along a river. The article did not state whether these rocks had been placed by humans or naturally occurred there. Animals do die of accidents in nature, but the risks increase greatly when humans enter the picture. When building structures in areas frequented by alligators and other wildlife, care should be exercised in the planning phase that minimizes risks to wildlife, including where roads are routed.

a

b

Figure 6.6. (a, b) An alligator at Brazos Bend State Park tried to enter the water via a pier and got stuck between the rails. Unable to free itself and keep its head above water, it drowned before anyone found it. (Photos by Dennis Jones)

Alligator Attacks

I have been able to find one, possibly two, fatal alligator attacks that occurred in Texas, as well as one serious attack where the individual survived. One fatality occurred in conjunction with the seventeenth-century La Salle expeditions to Texas. Dumesnil, the chamber valet of La Salle, was swimming across what is believed to be the Colorado River to see if the other side was firm enough for the horses. The man was attacked at his shoulders and pulled under. The writer Henri Joutel noted that alligators normally fled from humans and that this was the only instance where one of their men was lost due to an alligator. He also observed that wherever the natives bathed, they made the alligators flee (Foster 1998).

A fatality probably due to an alligator was associated with the Runaway Scrape of 1836. A wealthy and prominent gentleman from Nashville, Tennessee, Grey B. King, his family, and three slaves were headed toward thousands of acres of land that he had bought, mainly around the city of Austin. Their timing was not good. When they arrived in Texas, Santa Anna's army moved east from San Antonio, causing thousands of settlers to flee before it. General Sam Houston wanted scattered families from all parts of Texas to gather in one location to provide safety for the wives and children, while the men and older sons helped the Texas army. General Houston appointed Grey King to lead the families of the soldiers to safety. In April 1836 King waded in Buffalo Bayou at what is now called the San Jacinto Battleground and was shoulder deep in water when an alligator was sighted beside him. He went under and was never seen again. It was presumed that the alligator got him. There are several documentations of this event; although there are some minor differences in the renditions of the story, the basic details of the attack are pretty much the same (Anon. 1901; Griffin 1931; Hardin 1991).

There was also a nonfatal attack associated with the Runaway Scrape, which Mary Margaret Kerr related to her daughter. At the time Mary Margaret was a young girl fleeing with other families to safety. Their group was camped overnight on the banks of a bayou and awoke to screams and cries for help, as an alligator dragged a male slave by the leg toward the water. Someone grabbed a flaming stick from the campfire and thrust it into the alligator's eyes. The alligator released its victim and went back into the water (Crain 1957).

Statistics

Alligators certainly cannot be considered "human-eaters," yet there have been some run-ins with alligators that have resulted in severe injuries or death. There have been no recorded fatalities or headline-grabbing articles about someone losing his or her life or a body part to an alligator in Texas in the latter half of the nineteenth or in the twentieth century. Most attacks in Texas have been on duck hunters wading in the water, who stepped on an alligator. There was also an episode that Amos Cooper related to me involving children on a swing jumping into the water on the back of a submerged alligator that they knew was somewhere in the water. Amos also told me

about a child getting scratched on the arm by an alligator when the youngster got too close to the alligator's nest. TPWD's website section "About Alligators" states that "'attack' reports in Texas are usually more accurately described as 'encounters.'" We do hear on the local news from time to time about dog attacks that have resulted in severe injuries or death to a human being, yet most of us do not think of dogs in general as killers. With that in mind, I examine the statistics for alligator attacks and fatalities in the United States.

Six of the 10 states where alligators range have reported alligator attacks, according to statistics gathered from the various state wildlife agencies and research universities, compiled by the International Shark Attack File, and posted on the Florida Museum of Natural History's website. Here are the data for each state: Alabama, 5 ([date unknown]–2003); Florida, 326 (1948–2003); Georgia, 8 (1980–2001); Louisiana, 2 (1976–2003); South Carolina, 9 (1976–2003); and Texas, 15 (1980–2003). For the same periods, Florida recorded 459 shark attacks, followed by South Carolina with 23, Texas with 11, Alabama with 4, and Georgia and Louisiana each with 3. The only state to have fatalities from either animal was Florida, which had 13 alligator and 8 shark fatalities through 2003.

Why the disparity in alligator and shark attacks and fatalities in Florida compared to those in other states? First, Florida is a peninsula and juts out into the ocean. That fact, coupled with a large population, both resident and tourist, is responsible for the shark attack and fatality data. The alligator data are also a result of the large human population, combined with the large number of water bodies found throughout the state and their position relative to housing developments. Louisiana is almost all alligator habitat but lacks the large human populations where the alligators are found to produce the problem situations. Texas is a large state but possesses substantial alligator numbers only in the southeastern portion. The highest numbers of alligators live in areas like Anahuac. These high-alligator but low-human population areas tend to have another factor in their favor against alligator attacks. The people in these areas have generally grown up living with the surrounding land and its wildlife as an integral part of their lives, tending to make them more savvy about nature in general.

TPWD reports that most of its 400-plus nuisance alligator calls in Southeast Texas in 2003 were not true problem alligators, that more than 100 alligators were relocated mainly from subdivisions adjacent to natural habitat, and that a like number were killed in similar situations. Thus, the "hot spot" for alligator problems in Texas is the suburbs.

Circumstances of Some Florida Fatalities
I have no doubt that sometime there will be a severe injury or fatality due to an alligator in Texas, and it will be primarily due to substantial subdivisions being created in areas that were formerly not inhabited by humans to any major extent, particularly in the Houston area. For this reason, I take a look at the 16 fatalities due to alligators that occurred in Florida from August 1973 through March 2005 to see if there are any trends that would be useful in

predicting under what circumstances alligator attacks are most likely to occur and what precautions to take to avoid them.

The parameters that I looked at regarding fatal attacks were time of day, month and year, activity victim was engaged in at the time of attack, whether or not a dog was present or contributed to the attack, if the alligator was used to being fed by human beings, age and sex of the victim, and sex and size of the alligator.

Nine took place at various times during the daylight hours, and four occurred at dusk or night, all during March through October, when alligators are the most active and are feeding. The most common month was June with five fatal attacks, which is at the height of nesting. In each of the five fatalities between 1973 and 1987 the individuals were swimming when the attack occurred (one was even snorkeling). Individuals ranged from 11 to 52 years old, including a 16-year-old female and a 29-year-old male, both in the prime of their lives. Five attacks are not a statistically tenable sample size; however, the attacks occurred over 14 years, which is a rather substantial period of time. The animals were probably not habituated to humans, and swimming in areas where alligators occur may lead to attacks, as the prone position makes humans easy to grab and overpower.

The attacks in 1988 to 2005 were on people who were wading (4), swimming (2), or walking (1) on a narrow strip of land between two wetlands with a dog. In four other instances, circumstances regarding the attacks were unknown and the individuals were found after the attacks had occurred. Again the sample size is not statistically tenable. These attacks are probably indicative of a higher concentration of alligator and human populations in the same space, with the alligators losing their fear of people, so they began to attack people wading in water, not just swimming, where they were easier prey. The four individuals who were wading were bending over in the water, making them appear smaller and easier to pull over.

In three cases dogs were thought to be a contributing factor, and in another three the alligators were known to have been regularly fed by people. In four instances the individual ranged from 2 to 10 years of age, and in three cases the individual was 70 years or older. Children and elderly people are known to have slower reaction times. They may also be perceived by alligators as being easier to "take," the children being smaller than adults and in some cases prey size, and the older adults moving more slowly and feebly than their younger counterparts. Sex of the individual did not seem to make a difference if swimmers, children, and the elderly are excluded, which leaves two victims, a man and a woman, both in their 50s.

Ten of the alligators were 10 feet (3 m) or longer. Three alligators were 8 feet (2.4 m) or longer, and one was just under 8 feet. The other two alligators were 7 feet (2.1 m) and 6.5 feet (1.9 m). The 7-foot alligator was the only one identified as a female and killed a 52-year-old swimmer. The 6.5-foot alligator killed a 2-year-old child. Thus, larger alligators tend to be the most likely to be involved in lethal attacks, but smaller ones may be dangerous to children or an adult who is in a totally prone position. Since few female alligators reach a length of 10 feet, most alligators involved in lethal attacks tend to be males. The largest alligator causing a fatality (12 feet, 4 inches, or 3.8 m) was known to be old and in poor health. When it killed a swimmer, it may have been trying to get easy prey.

The fatal attack in 1988 had all the elements of what can contribute to an alligator attack, especially one in which serious injury or death is involved. It was in the month of June, the alligator was over 10 feet in length and had been fed by humans, the victim was 4 years old, and she was wading in the water with a dog. Furthermore, the girl was throwing rocks and kicking the water and was believed to have kicked the alligator. The alligator bit her in the stomach and then dropped her. As she began to crawl away, the alligator grabbed her and carried her into the water. I believe that the alligator started out trying to get the girl out of the water, away from him, by biting and letting go. If the victim had been an adult, the bite would have probably been on the leg, and the adult would most likely have still been on his or her feet and able to move fairly quickly away from the water. Due to the child's small stature, the alligator bit the child higher up on the body where the vital organs were, resulting in very serious wounds and knocking her off her feet. The child was then in a prone position and moving very slowly. At this point, the girl went from being an intruder to being a potential prey item.

How To Survive an Alligator Attack

The best strategy to avoid being attacked is not to put yourself in a situation where your vulnerability to an attack is enhanced (or even possible). That being said, the best way to decide what methods to use to try to get away from an alligator that has seized you is to examine what methods worked for survivors of serious attacks. The goal is to get the alligator to let go and then immediately get to safety where it cannot pursue you. Punching the alligator in the face or poking it in the eyes seems to be the most effective methods of getting an alligator to release a victim. Punching the alligator worked in the case of two swimmers, and the 1836 alligator attack in Texas was halted by the alligator being poked in the eyes with a fiery stick. A woman in Asia was able to get a 15-foot (4.5 m) crocodile, which had her in a death roll, to release her after she poked it in the eye with an iron hook.

If you are with someone who has been seized by an alligator, pulling the person away from the alligator with all of your strength may work. Friends were able to save a man from a 12-foot (3.6 m) alligator that had grabbed him by the leg and was slinging him around by doing just this. A man saved his dog from a 6-foot (1.8 m) alligator by "wrestling it" after the dog had jumped out of the boat onto the alligator and subsequently was attacked.

Addendum

As this book goes to press, two alligator attacks have occurred in Texas, one fatal. The attacks took place in core counties within days of each other, and both alligators were believed to have been displaced from their homes by recent flooding. Both victims were swimming at the time of the attacks. The nonfatal attack occurred on June 28, 2015, when a 13-year-old male got overheated on a family fishing trip to a park in Chambers County and decided to go for a swim. The victim was in waist-high water and felt something grab his arm. He yelled that something had grabbed him, and both of his parents and his grandfather jumped in the water and successfully pulled him away from an alligator that was estimated to be 6 to 8 feet in length. However, the father became a second victim when the alligator bit him in the leg and held on. The father managed to free himself by kicking the alligator. The son's injuries involved one arm and one leg. Both father and son were hospitalized and recovered.

The fatal attack occurred at a marina in Orange County on July 3, 2015, and involved a 28-year-old man who had recently moved to the area from a northern state, where there are no alligators. He decided to go for a late-night swim, despite the "No Swimming, Alligators" signs posted nearby and verbal warnings from employees at the marina's restaurant. He said "blank the alligators" and jumped in the water. He was almost immediately grabbed by a large alligator. His body was seen a short time later floating near the pier and was recovered before morning. The alligator was killed and measured 11 feet, 6 inches (3.5 m), presumably a male due to its size.

Several smaller alligators frequented the pier area and were fed by visitors to the marina. The large alligator involved in the attack had only recently appeared in the area and allegedly hung around the pier and was fed. Since the attack occurred right after the man jumped in the water, it was postulated that the alligator was under the pier and rushed forward for an expected feeding. This tragedy indicates that alligators can become habituated to humans in a short period of time, especially if they become displaced from their normal home range.

Solutions to Alligator Problems

Solutions to problems with alligators involve education about alligators and their habits, knowledge of laws concerning alligators, and ways to minimize human-alligator contact.

Education

Educational signage at parks and community lakes can provide information about alligator life history and habits, as well as how to behave around them. Such signage does not require personnel to be actively giving lectures to teach the public about alligators and is available 24 hours per day. However, education sessions conducted by a knowledgeable individual certainly can disseminate a greater amount of information than a sign, and questions can be answered as well.

TPWD alligator biologist Amos Cooper says that he and other TPWD personnel provide talks, presentations, and a chance to see and touch juvenile alligators to schools, civic groups, nature clubs like Audubon, and at events such as Gatorfest, Cajunfest, and the TPWD Wildlife Expo. Game warden Barry Eversole has a talk with officials at new hous-

ing developments in Fort Bend County about alligators and sometimes gives presentations to residents of homeowners' associations at a monthly meeting. The interpretive naturalists at Brazos Bend State Park conduct presentations on alligators, and naturalists at other natural areas and parks in Texas may conduct similar programs.

In Florida, community centers have hosted alligator education sessions, and the same type of learning modules could be conducted at public libraries or other facilities in Texas to reach rural residents or those who do not live where there are homeowners' associations. Ideally, every citizen who lives in an area where alligators occur should be given the chance to learn more about alligators and how to coexist with them safely through free sessions offered at least annually. These courses should be taught by a TPWD official who is experienced with alligators or by a local animal control officer who has received training from TPWD. Both Amos Cooper and Barry Eversole think that a brochure on alligators handed out by realtors would be extremely helpful to educate new homeowners moving into areas where alligators reside.

Game wardens are the main point of contact for alligator problems and public education. They are the individuals initially called on regarding nuisance alligators, even if a nuisance alligator hunter is the person who ends up removing the alligator. TPWD now provides training for each game warden cadet class in Texas that includes hands-on experience with several different sizes of alligators, guidelines on their humane treatment, and how to work with the public.

Law Enforcement

Enforcement of laws on the books can coun-
teract many behaviors of the public that are
both illegal and dangerous. For example, having "dogs at large" (off leash and not penned)
or "feeding alligators" are both acts that could
lead to a human being attacked or killed by an
alligator. These need to be enforced with maximum penalties imposed. Signs should be posted
stating that feeding of alligators is prohibited
and listing the penalties.

Ascertaining the Presence of Alligators

The first priority in an area where alligators are
a potential is to ascertain if they are present or
not. It is not unusual to be totally unaware of
their presence until someone points them out.
In fact, there are many areas where alligators
exist, are not habituated to humans, and therefore go unnoticed by the general populace. The
problem occurs when they are discovered and
then one portion of the population gets hyster-
ical and the other portion feeds them, creating
a problem. The easiest way to ascertain if alligators exist in an area is to go to the local water
bodies at night and shine a bright light to catch
the red eyeshine of the alligators. Frogs also
have eyeshine, but it will be greenish in color.

Open trails in the water of thickly vegetated ponds and lakes or openings in shoreline
vegetation are other telltale signs. The use of
binoculars focused on these areas at various
times of the day can help ascertain whether an
alligator or some other type of animal is making these trails. When the ground is wet, it is
possible to see alligator footprints, as well as
tail and body drags in the soft mud. During the

late winter through spring and again in the fall,
alligators can be seen basking on the shore or
on logs or islands in the water bodies that they
inhabit.

The presence of alligators in a lake or pond
near human habitation is not automatically a
problem. Yes, the alligator will eat anything in
the lake or at the shoreline that is small enough
to eat. Dogs and cats are supposed to be on
leashes (i.e., leash laws). Cats are unlikely to be
eaten by alligators. They are more likely to fall
prey to other predators such as dogs, coyotes,
and bobcats. Alligators can actually provide a
"popular viewing experience" for residents,
just as watching birds does. The big difference
is that you do not want to feed your viewing
subject.

*Posting Signs, Increasing Visibility,
and Controlling Habitat*

Signs should be and sometimes are posted in
populated areas where alligators are known to
occur, warning of their presence, not to feed
them, and not to enter the water. Occasionally unscrupulous realtors "pull" such warning
signs where they are trying to sell homes.

Areas around water bodies should be kept
well mowed so that there are no surprises to
alligators or humans that accidentally wan-
der into each other's path. Vegetation such as
bushes and trees that obscure visibility should
be located only in the areas surrounding water
bodies that are low-traffic areas for humans.
Marshy, swampy areas that are not a part of a
regular shoreline should be filled in or regraded
if they are in an area that gets high usage by
humans, as long as increasing the elevation of

such areas does not violate wetlands protection regulations.

In the case of sizable ponds, basking islands can be built well away from the shoreline to provide areas where alligators can bask and nest without being disturbed by humans. A key factor to the success of these islands is management of vegetation on the islands. It should provide some shade in the form of trees and shrubs and contain ground cover, such as grasses, that are useful in nest building. It is also important that these islands be kept open to maximize usage by alligators and other wildlife. Every few years it may be necessary to clear dense vegetation, such as blackberries and other weed species, which threaten to overtake the islands. This is best accomplished during the winter when the alligators are relatively inactive.

Another technique that is useful to minimize the possibility of humans and alligators interacting is to locate/relocate areas that provide potential alligator habitat away from high human usage. However, in the case of housing developments this may pose a problem concerning detention ponds required to prevent flooding. As the countryside becomes more a "cement city," it is necessary to provide such areas to prevent flooding when heavy rains produce substantial runoff that can no longer soak into the ground. Housing developments are being built with detention ponds as decorative lakes that contain water all of the time, instead of just in periods of heavy rains, to be used as a marketing tool to sell homes. This has two adverse effects. One is that the pond is already holding water so will not be able to accept as much additional water during heavy rains or a major hurricane. The other is that it is actually creating habitat for alligators in an area of high human population and usage.

Rural homeowners who fear alligators may want to reconsider building a pond on their property, at least near the house and adjacent yard area. Ponds also attract a variety of wildlife in addition to alligators. Game warden Barry Eversole was contacted by a woman who wanted a subadult alligator removed from her pond. He went down to the shoreline of the pond, which had tall vegetation, and noticed that there were a number of venomous cottonmouth snakes, which were undoubtedly providing a steady food source for the alligator.

"Doggy doors" may also pose a problem. For example, a family had recently installed a pet door for their small dogs and cat. No one had been home since very early in the morning. When the woman and her son returned to the house at about 8:00 p.m., they were greeted by a 5-foot alligator, making itself at home on the kitchen floor. The dogs and cat had gone upstairs. Evidently, the first cold weather had hit and the temperature had dropped a substantial amount in just a short period of time. The alligator was just trying to get out of the cold! Barry Eversole told the woman that she was very lucky that she didn't have a couple of raccoons come into her house. They would have been very destructive, could have gone upstairs and wreaked more havoc, and the family pets would have not been able to get away from them. Needless to say, the family had the pet door removed.

Erecting Barriers

Erecting a fence or other barrier that limits the access to shoreline areas where alligators bask or nest can alleviate potentially negative alligator/human encounters. For example, a permanent fence with four to five boards closely spaced can serve as an attractive deterrent to partition off swampy areas of a water body or locations known to harbor basking or nesting alligators on a regular basis. For sudden or seasonal problems with alligators "pulling up or basking" onshore or building a nest, the easy fix is orange safety barricade fencing (fig. 6.7), which is available from most home improvement stores. It is durable, inexpensive, quick and easy to erect with T-posts, and it can be

taken down when it is no longer needed or replaced by more attractive, permanent fencing. A big advantage is that it can be extended out into shallow water, minimizing the risk of more adventurous humans trying to invade the blocked-off area.

Using Tools for Safety and Piers for Observation and Fishing

Several of the alligator fatalities in Florida involved individuals bending over to weed, pick plants, or similar activities in the shallow water. When in the field, I always take a snake field hook, not just to turn over logs to look for reptiles and amphibians or to pick up snakes but also to serve as an extension of my hands

Figure 6.7. Temporary barricades constructed of orange safety fencing can keep humans from approaching too close to an alligator on land.

and feet to help keep me safe. Before reaching or stepping out into the water where alligators may be in quiet repose before being splashed, kicked at, or stepped on by a human intruder, you can use a field hook or even a hiker's walking stick as a "feeler" for the presence of an alligator, giving both human and alligator sufficient time and space to react to the presence of each other without having direct contact. If this indirect contact occurs, then one or both back off. Observation or fishing piers can serve as buffer areas and protect humans from direct contact with alligators.

Steps to Take against Habituated or Aggressive Alligators

Whenever alligators become aggressive, take the following steps in order as necessary to deal with the problem: (1) post signs and erect barriers; (2) educate the affected public and strictly enforce laws; (3) conduct aversion training; (4) relocate an alligator or nest; and (5) remove an alligator to a farm or euthanize it. Note that in Texas steps 3–5 should be carried out only by TPWD and are detailed in the following discussion.

Conduct Aversion Training
Whenever alligators first cross the line and react to humans feeding them or conducting any activities that can cause them to lose their natural fear of human beings, aversion measures may serve to, at least temporarily, halt their aggressive responses. Literally hitting them with mops in the face or spraying them with a "bear spray" used to repel bears if they come toward the shoreline in response to seeing people is one way of providing negative reinforcement. Another method is to noose the alligator, "exercise," and release it. Exercising the alligator involves pulling it up close with the rope, letting it swim away, and alternating the up-and-down actions with pulls from side to side. The process is continued until the animal is exhausted. Care must be taken to release the animal onshore rather than in the water so that it does not drown from stress and exhaustion. The aversion training is usually successful on a short-term basis but may have to be repeated within a few months if the alligator resumes its problem behavior. A way to train alligators to associate humans feeding them in a negative context is to feed them raw ground meat laced with habanero peppers.

Relocate an Alligator or Nest
Moving an alligator to another location is an option that needs to be thoroughly thought out before being executed. It is in reality an extreme measure that has potential deleterious side effects for the alligator and humans involved. First, alligators are very territorial animals and often will try to return to their territory if they are 7 feet (2.1 m) or larger. If they are taken a certain distance away, often the distance in road miles is much longer than "as the crow flies," which is more likely the path an alligator might take rather than roads. The attempt to return places the alligator in unfamiliar terrain and potentially dangerous situations, particularly where it has to cross busy roads. The alligator may also come into contact with people during its journey, and this

Capturing an Alligator Using a Noose

Materials

- Two lengths of rope, each a minimum of 30 feet (9 m). The rope should not be sisal or another type of rough rope that will chafe or abrade the alligator, nor should it be polypropylene or any type of rope that floats (a rope that sinks is needed to enable water captures). A 3/8-inch-diameter (0.9 cm) double-braid polyester line works well. A metal snare can be tied to the rope and used in place of a rope noose with smaller alligators.
- Catchpole 6 feet (1.8 m) long made of fiberglass, aluminum, or wood. A telescoping fiberglass pole used for replacing light bulbs is probably best. Wood is durable and long lasting, whereas aluminum is easily bent or damaged when bitten.
- Masking tape (for taping noose to catchpole).
- Duct tape or bicycle inner tube (for taping alligator's mouth shut). Caution should be used with inner tubes or rubber bands (used on juvenile alligators), as they can cut off circulation and leave marks if left on too long or put on too tightly. Masking tape is strong enough to tape juvenile alligators if it is wrapped around enough times.
- Cloth for covering the eyes of the alligator.

Setting Up the Noose and Attaching It to the Catchpole

To form a noose from the rope, tie a small bowline knot, forming a loop about 2 to 3 inches (5.08 to 7.62 cm) in length. Pass the other end of the rope through the bowline to form a larger loop that is used as the noose. A figure-eight knot can be tied toward the free end of the rope as a stopper knot, which helps prevent a rope from slipping from your hands.

The rope noose or metal snare with a rope attached may be used to capture an alligator when loosely attached to the end of a long catchpole. After the noose/snare is placed over the alligator's head, the rope must be able to "break away" easily from the catchpole. This can be accomplished by using masking tape to loosely fasten the noose to the catchpole. For conducting sideways captures, place the noose parallel to the catchpole and tape at two locations to give the noose "width." For head-on captures, place the noose perpendicular to the end of the pole and tape only once. After the alligator is noosed and the rope is tightened around the alligator's neck, quickly set aside the catchpole and hold the rope tightly.

CAUTION: Once an alligator is noosed and tension is placed on the rope around its neck, the alligator will begin to roll in an attempt to throw the rope. Caution must be exercised that the rope is not allowed to wrap around the animal and shorten the distance between you and the animal. The alligator will struggle against the rope and always back up against the opposite pull of the rope. If you slacken the

A rope noose or metal snare can be loosely secured to a
catchpole with masking tape, which allows the noose to
"break away" from the catchpole.

rope around your hand or place your hand in
any loops in the rope.

Ropes fastened to the bumpers of trucks
can be used to haul large animals onto land. In
doing so, it is important to go very slowly and
have an onlooker make sure that the animal is
not unduly stressed or that there is no rough
surface or vegetation that the alligator could
be pulled over that would cause abrasions (a
muddy or grassy bank is ideal). The fat around
the neck of large animals should prevent the
rope from cutting off the airway (i.e., closing
off the trachea). Very little force is needed to
pull the alligator compared to its weight, especially if it is floating in the water.

Securing Very Large Alligators

After the alligator is noosed and somewhat
fatigued, a second noose rope is placed around
the alligator's neck. This is accomplished by
pulling the second rope under its lower jaw and
neck, tossing one end of the rope over the animal, and passing the rope through the loop tied
in one end to make a noose. The two ropes are
pulled in opposite directions to limit the movement of the animal. The ropes are then secured
tightly to stationary objects.

If Only One Jaw Is Noosed

Occasionally an alligator opens its mouth just
as the noose is going over its head, and only the
upper jaw is noosed. This is a difficult situation
in that the alligator is captured but not in the
intended manner. The best course of action is

tension on the rope, the animal may move forward toward you. However, in most cases the
alligator will hold its ground or try to run away.
If the crocodilian is large, you may need some
stationary object to fasten (belay) the rope onto
so that you cannot be pulled into the water or
dragged along the ground. Do not wrap the

If the upper jaw is noosed, the best course of action is to throw a second noose around the alligator. The first noose can then be safely removed with a field snake hook.

to throw a second noose around the alligator as described for securing very large alligators. A field snake hook can then be used to loosen and remove the first rope from the jaw.

Securing the Alligator

Adult and large subadult alligators should be put into a catchpipe for immobilization and their mouths secured if need be. Juvenile alligators and small subadults can be grabbed

by hand behind the head, the snare or rope removed, and the mouth held shut in order to tape it.

Releasing the Alligator

Two notes of caution need to be addressed before releasing the alligator. It should never be released directly in the water after the stress of being handled. Large adults especially are prone to drowning under these circumstances. See the catch-pipe technique for releasing the adults and large subadults. Care should be taken that juvenile and small subadult alligators do not escape to the water before removing the tape or rubber bands from their mouths. It is advisable to have a second person cut the tape or remove the rubber band while the animal is being securely held to prevent such escapes and minimize the possibility of the handler getting bitten.

An alligator being released at the water's edge from a catch pipe. It is not unusual for the alligator to roll and twist around as it is being released, so care should be taken to protect it from sharp or hard objects while it backs out of the pipe.

can be an unsafe situation for both. Subadult alligators moved to areas where the population levels of alligators are already high or there are a substantial number of large adults are at risk of being cannibalized or chased away from the site.

If a determination is made to relocate an alligator, then a decision should be made about how it will be captured: using a trap or noosing it around the neck. If trapped, it can be transported inside the trap. If noosed, then the catch-pipe technique can be utilized to safely restrain and transport the alligator with a minimum amount of stress.

When an alligator nests in an area that is close to people and it is not feasible to set up a barricade, it may be advisable to move the nest. In rare circumstances, it may be possible to move the female and her entire nest to a new area that would be suitable for an alligator nursery with few to no alligators to bother her. However, once a female is captured and taken away from her nest for a short period of time, she may lose interest in it. In such situations, it is recommended that it be checked the next day to see if she is tending and guarding it.

Ideally, if an incubator is available, the eggs could be carefully removed from the nest, each one retaining its original orientation. If it is early in the incubation period when the sex is not yet determined, it is possible to pick a temperature to get a mix of both sexes or predominantly one sex. Keep in mind that the temperature in the incubator may be inaccurate, and opening the door may drop the temperature for a period of time, thus affecting the true incubation temperature.

If it is possible to take the entire nest or most of the nest intact (grass nests are more conducive to this than ones with a lot of soil and sticks), then it could be relocated outside to an area protected from predators (perhaps placed within a raccoon-proof chain-link enclosure with a chain-link top as well). The eggs could be removed as they begin to hatch. It also is possible to place the nest inside an outdoor building, such as a garage, which passively heats and cools much the same as the outdoor temperature fluctuates at the original nest site. The eggs can be brought inside a house or other facility where temperatures are kept between 72°F and 75°F (22.2°C and 23.8°C), but there must be a heat lamp or other means (something that does not pose a fire hazard) to keep the nest warmed to the proper temperature range.

At Brazos Bend State Park, the babies hatching from a nest that is artificially incubated are released with a female alligator that is known to be a good mother and currently has young. On one occasion, volunteers at the park rescued eggs from a nest that was close to hatching, because it was in an area that was flooding. I suggested that the hatchlings be returned to the nest site so that their mother could tend to them. If no female were present suitable to raise the young alligators, I would recommend that they go to an alligator farmer for rearing. If the nest is not in a park or wildlife management area, it would be advisable to give the entire clutch of eggs, at the time the nest is removed, to a farmer involved in the egg-collection program as part of his annual quota so that they can be kept in a incubator and hatched under optimal conditions.

Catch-Pipe Technique for Restraining and Transporting Alligators and Other Crocodilians

Dennis Jones suggested that we could restrain alligators with much less stress to the animal or the handlers if we placed the alligators inside a large PVC pipe after they were noosed. Dennis and I then worked out the bugs by constructing a prototype and using it on alligators that we captured. The pipe was modified to suit our needs within a short period of time. The techniques and protocol for utilizing it evolved over a period of several months during which we made multiple captures of adult alligators (see Jones and Hayes-Odum 1994). The procedure can be accomplished with two people. However, additional personnel may be needed if the alligator needs to be lifted into a vehicle for transport. The pipe should be stored out of direct sunlight, as PVC pipe deteriorates from extended exposure to UV light.

The diameter and length of pipe that we used were suitable for adult alligators up to 10 feet (3 m) in total length. If a smaller species or individuals were to be handled, it would be desirable to have a smaller diameter and length of pipe to make it lighter in weight and easier to handle and to be sure the animal fits more snugly inside the pipe. Conversely, a larger diameter and length would be needed with extremely large crocodilians. Saumure et al. (2002) took our concept of restraining crocodilians in a PVC pipe a step further by coming up with a method to use a PVC pipe to capture (rather than use a noose) as well as restrain crocodilians in captivity. I want to emphasize that as a capture/trapping technique a PVC pipe is effective only in an exhibit-type situation and that Barry Eversole's trap is a good choice for capturing wild alligators.

Materials

- White PVC pipe (50-pound plastic irrigation pipe) with an inside diameter measuring 12 inches (0.31 m), a wall thickness of 1/8 inch (0.32 cm), and cut to a length of 10 feet (3 m). Holes of a diameter sufficient to allow the passage of a rope without binding should be drilled along the length of the pipe in a straight line at intervals of 6 inches (15.2 cm).
- Noose rope and its associated materials (see specifications in "Capturing an Alligator Using a Noose").
- Rope (9 feet, or 30 m long) for inguinal region (area immediately in front of back legs) of a material other than polypropylene or a rough substance such as sisal to avoid chafing the animal.

Getting the Alligator into the Pipe (Illustrations by Lisa McDonald)

a) The procedure starts with the alligator being noosed around the neck with a rope or after being captured by some kind of a trap

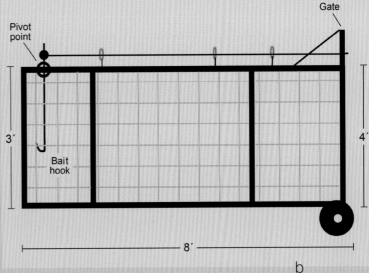

Trap 8' x 3' x 3'
Frame 1" x 1/8" welded angle iron
Bait hook
Trigger rod 3/8" rebar
Gate door 4' high
Gate track slide 1 1/4" angle iron
Horse panel (5' x 16') enclosure; panels have a 1" x 3"
square grid to have the ability to trap very small alligators.
Wheels on the gate end of the trap help maneuver and guide
trap into water. Gate opening should not be submerged
more than half way in water.

(a) Game warden Barry Eversole with a trap that he built to capture alligators. Traps like these are being used by TPWD to capture nuisance alligators rather than noose them. (b) Alligator trap materials and specifications. (Illustration by Lisa McDonald)

or line with a wooden dowel. The end of the noose rope is passed through the pipe. We found that the quickest and most effective means to accomplish this is to have a permanent knot with a small loop at the opposite end of the rope that secures the alligator's head. The hook portion of a snake field hook is then hooked into the loop, and the snake

hook is thrown into the pipe handle first. The weight of the hook ensures that the end of the rope will quickly and completely go through to the opposite end of the pipe. The snake hook is then removed, and the rope sticking out the end of the pipe is held to ensure that the alligator does not pull away with the rope.

b)   The pipe is guided over the animal's head. This is the only difficult stage of the technique, as some alligators begin to roll after they are noosed, which makes it harder to get them started into the pipe. When the head is inside the pipe, the pipe is held in position by applying constant tension to the rope. The pipe should be rotated so that the holes are on the top. The forelimbs are folded back against the body (so that the toes point to the tail end) by the pipe, as the animal is pulled into its lumen (inside of pipe) until the back legs come even with its end. (Note that in the drawing the animal depicted is smaller and the legs did not stay folded back as they would with a larger animal.) A second rope is then tied immediately anterior to the hind limbs (the inguinal region).

c)   By pulling on the anterior or posterior rope, we can position the animal within the pipe so that the head or caudal (tail) region extends a desired distance from an end of the pipe. If the head region is to be accessed, the jaws can be tied shut with a rubber bicycle inner tube or duct tape to keep the jaws closed. While the animal is in the pipe, it is not necessary to cover its eyes. However, if we position the head outside the pipe, we then drape a cloth over the crocodilian's eyes to keep it calm.

d)   When the animal is in position, the rope at each end can be secured by tying it in a knot at a nearby hole in the pipe. The tension on the ropes should restrict forward or backward movement within the pipe but allow enough freedom so that the animal can move its head up and down to prevent regurgitated stomach contents from blocking the glottis.

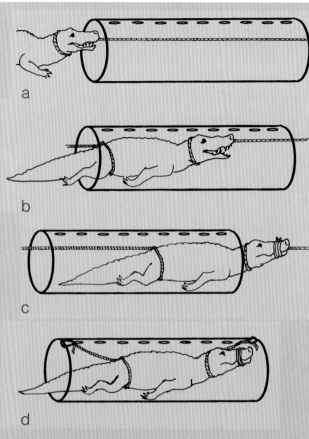

Measuring the Alligator

The regularly spaced holes in the pipe can serve as an aid in obtaining measurements of total and snout-vent lengths, as well as providing places to secure the ropes. The holes allow the snout and tail tips to be seen; thus, a tape measure can be placed on top of the pipe to get an accurate measurement of total length. Moreover, if the tip of the snout is positioned at the exact end of the pipe, much of the measurement can be calculated without the use of a tape measure, by knowing and/or marking on the pipe the distance between holes. Snout-vent length can be determined by lining up the end of the cloaca with the end of the pipe and measuring forward.

Releasing the Alligator (Illustrations by Lisa McDonald)

a)   To release the animal, the rope at the tail end is removed first.

b)   The animal is then pulled to the front of the pipe, until the rope on the neck can be released with the aid of a snake hook. The inner tube is pulled off with a snake hook rather than untying it. If duct tape is used, it should be carefully cut with a knife or scissors at the side of the mouth and removed nearly simultaneously as the rope at the neck. CAUTION: The eyes should be covered with a cloth before removing the inner tube or tape from the mouth and the rope from around the neck.

c)   The pipe is tipped upward and/or jerked backward/upward until the animal is induced to leave it. CAUTION: Never release an alligator directly into the water after its capture. Always release it on land and allow it to return to the water under its own power. Returning an alligator into water after stressing it during a capture can cause it to drown.

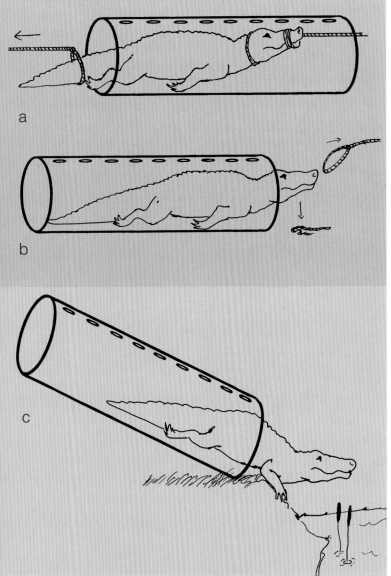

Euthanasia

When a crocodilian is euthanized in a zoo, it is generally accomplished via an intramuscular (IM) or intracoelomic (i.e., the body cavity) injection of phenobarbital combined with some other drugs. Such drugs are highly regulated and therefore not easily available for nuisance alligator situations. Additionally, in the case of farmed or wild harvested alligators, the use of drugs would not be appropriate due to the resulting meat products being used for human consumption.

The most common way to kill a crocodilian in nuisance, hunting, or farming situations is by shooting it with a gun, or by using a knife or arrow, all in an effort to hit the brain and cause brain death. Brain death results in the animal not being able to feel pain or to consciously make any movements. It does not eliminate reflex actions.

The "Code of Practice: Taking, Handling and Transportation of Crocodiles" (which can be found online) drawn up by the Australian Environmental Protection Agency, covers the humane treatment of captive and wild Australian crocodiles and includes a section on euthanasia. In addition to euthanasia drugs, the document lists shooting through the back or side of the cranial platform or between the eyes. It recommends using a rifle or shotgun, unless the animal is firmly secured, in which case a pistol can be utilized, and gives specifications for the firearms. Another method detailed for crocs under 6.5 feet (2 m) in length is a hammer blow applied to a sharp metal chisel just behind the skull, severing the spinal cord. Brain death is then accomplished by pithing the brain by insertion of a rod. Although the recommendations relate to crocodiles specifically, this protocol for euthanasia should be appropriate for other crocodilians of a similar size, with differences in skull anatomy in all probability playing a negligible role in the final outcome.

I cannot recommend that any crocodilian be killed with a firearm unless it is confined or restrained. A trial hunt by Texas Parks and Wildlife was undertaken during September 6–18, 1985, to evaluate the ability of alligator hunters to shoot free-swimming alligators with firearms (Johnson and Thompson 1986). Of the 117 alligators shot at, 37 were missed, 43 were killed and retrieved, 17 were visibly hit but not retrieved, 20 were possibly hit, and 1 was later found dead of injuries directly related to the hunt. The conclusion of the study was that if firearms were the only method utilized for alligator hunting, the annual harvest-related mortality could be as much as 46.3 % more than the harvest quota. In assessing the shooting of free-ranging/free-swimming alligators as a method of humane killing, it falls short of the mark, as there is no guarantee that the animal will not die a slow, painful death or be severely injured and left alive with slow-healing wounds that cause permanent disabilities.

Another method is that used in slaughterhouses: compressed air–driven devices similar to those used to humanely kill cattle and other hoofstock. Zilca Campos, a researcher at the EMBRAPA-Pantanal (Empresa Brasileira de Pesquisa Agropecuaria) of Brazil, assisted in the development of such a device because of ethical concerns for diminishing the suffering of animals, fulfilling technical demands of international organizations in regard to humanitarian slaughter, and improving the quality of the meat. The Australian "Code of Practice" considers a captive bolt gun an acceptable method of humanely killing a crocodilian as long as the equipment is fully functional and personnel are skilled in its use.

Remove an Alligator to a Farm or Euthanize It

A nuisance alligator can be transported to an alligator farm, where it can be used as breeding stock or maintained until it is harvested. This is a likely occurrence when the nuisance alligator hunter is also an alligator farmer. Killing a nuisance alligator is sometimes the only viable solution, and the animal is killed either by a game warden or nuisance alligator hunter. Unfortunately, if a nuisance alligator is to be relocated or euthanized, euthanasia is at times a much better choice. It means that an alligator will die a humane death shortly after capture rather than suffer a high risk of death in many relocation situations where, for example, it may get hit by a car and not die immediately.

The methods utilized to capture the alligator and transport it for euthanasia are the same as for relocating an alligator. An alligator should not be killed at the capture site unless it has been seriously injured, such as being hit by a car. In such an instance it is most humane to euthanize it on-site as quickly as possible. Alligators are subject to feeling pain, as are all higher vertebrates. Therefore, it is important to realize that the use of immobilizing drugs such as gallamine triethiodide or succinylcholine chloride does not keep the animal from experiencing pain, only to be incapable of doing anything about it. While exposing an amphibian to extremely cool temperatures "slows it down," it does not have the same effect on alligators; therefore, the use of ice or any type of refrigeration does not serve any practical purpose and is inhumane as well. An effective euthanasia technique is one that is fast and effective, and where brain death is immediate.

Captive Alligators Released into the Wild

In most cases a captive alligator should never be released into the wild. Although sometimes an introduction seems to be successful, there are potential problems that pertain to the well-being of the alligator being released, the other alligators in the population that it is being released with, and the human population that may come into contact with this alligator.

Any alligator set free at a site with other alligators is at a disadvantage. It knows nothing about its new location and is faced with other alligators trying to chase it off, fight with it, or kill it, particularly if it is "dumped" into the territory of another alligator. Subadults are especially at risk in this situation. The alligator population as a whole could be infected by parasites, bacteria, or fungi that were introduced by this single animal. The confines of captivity tend to multiply such pathogens, whereas they may exist in the wild at extremely low levels or not at all. If the animal is released in a park or other area where it comes into contact with humans, it may show some behaviors of a habituated alligator.

If a park or other facility has exhibits and maintains young alligators for education programs, I do not have a problem with these animals being released under certain circumstances. The two most important points are that they need to be released when they are still young enough to be cared for by a female alligator who is an experienced mother, and she needs to have similar-sized young. I have successfully released hatchling, yearling, and two-year-old alligators with females that had the same-sized offspring (that is, alligators that were 2 feet or less in length). I have never released offspring to a female that were not the same size as her own, and I am unsure how she would react. However, if no females are available with the same-sized young, then the next best thing would be to release the younger or older offspring with a female that is protecting some young and check back to make sure that she is caring for them.

I released some 2-foot alligators with a female with the same-sized young that I had been keeping in a naturalistic enclosure outside for over a year. They were seizures that game wardens had made when they were hatchlings and had not had any contact with adult alligators for nearly two years. I had used them from time to time in educational programs, but other than those occasions they were pretty much wild alligators. However, when I released them with a female alligator with babies their age, they gave distress calls seemingly in response to the female alligator's presence. It seemed that they were imprinted on humans rather than a large alligator for protection. They looked on the female alligator as a foreigner. I checked back on those juveniles weekly. When they saw me, I would immediately know which ones were my previous boarders, as they gave juvenile distress cries in my direction when they saw me. They stayed with the other young alligators but avoided the female alligator in attendance. Finally, by my fourth visit, they had accepted her, and I heard no more juvenile distress calls on that and subsequent visits.

Alligators larger than 2 feet that have been kept for educational programs should never be allowed to be released into a wild situation. These captives are usually kept in stock tanks in relatively shallow water. They do not know how to dive or act in deep water. This is a must for wild alligators, and most of these animals, when released, probably do not survive long in the wild. They will most likely be killed by larger alligators. The potential for spreading pathogens may also increase the longer that they have been in captivity and the larger their size. A much better decision would be to release them into the wild at a much younger age or to place them on an alligator farm if they are older, where they will be protected, revert to the wild, and be harvested at a later date.

At present, it is illegal in Texas to release former captive alligators into the wild, even at a TPWD facility (although it is being done). This regulation was set up to protect wild alligators from diseases that may be transmitted from captive animals.

• • •

I included these steps for dealing with nuisance alligators for two reasons. First, when alligators are considered to be possible nuisance alligators, public sentiment can be extreme in favor of or against the alligator. By understanding what steps to be taken, in what order, and if they need to be taken, Texas residents can work as a team with TPWD in trying to find a solution for the specific alligator situation that results in the best decision being made for the alligator and the residents. Second, not all TPWD personnel who respond to nuisance alligator calls have the same amount of experience or the same level of judgment. Hopefully, a framework such as this one will help them think through decisions that they make regarding a proper resolution of the problem.

Alligator Watching

If you see only an alligator's head at the water's surface, you can still determine how long it is. The length from the eye to the nostril in inches is a close approximation of the total length in feet. Therefore, an alligator that measures 7 inches (17.7 cm) from the eye to nostril would be about 7 feet (2.1 m) in total length. The accuracy of this measurement decreases in alligators 10 feet (3.0 m) or larger, and this measurement tool does not work for other crocodilian species.

Where to See Alligators in the Wild

Since alligators are native to Texas, there are ample opportunities to view alligators in the wild. The places to visit alligators in the wild and in captivity all have websites that provide details, such as entrance fees, maps, hours, and other information that will be helpful in planning a trip. If traveling a distance to visit a certain facility, you might see if any of the other places discussed here are nearby or on the way. Some of these viewing opportunities can take up an entire day, while others can be enjoyed in the span of an hour or two. Here are some of the best places to go to see wild alligators. Happy gator watching!

State Lands with Alligator Populations

Brazos Bend State Park Located in Needville, less than an hour's drive from Houston, Brazos Bend has been nicknamed "Alligator State Park." This park has the premier alligator viewing opportunities due to a large population that exists in multiple water bodies that are interspersed with hiking trails. Whereas boats are needed to view most large groups of alligators, here one can literally walk up to an alligator on a trail or see alligators engaged in such behaviors as basking, feeding, fighting, mating, and interacting with their young.

Caddo Lake Located on the Texas/Louisiana border with half of the lake in each state, the 26,000-acre lake boasts beautiful cypress swamp wetlands and has American alligators as one of its most visible wildlife species. It is the largest natural lake in the South and the only natural lake in the state of Texas.

Gus Engeling Wildlife Management Area Located in northwest Anderson County in Tennessee Colony, the management area encompasses nearly 11,000 acres of prime wildlife habitat and includes alligators among the many species found there.

J. D. Murphree Wildlife Management Area Located in Port Arthur, this coastal marsh area is teeming with alligators. It has boat access only. There are drawings for who can participate in alligator hunts.

Mad Island Wildlife Management Area This wildlife management area and its surrounding area near Bay City is known as a bird lovers' paradise. Alligator hunting is available to a limited number of hunters through a drawing.

Orange Visitors Center This is one of 12 Texas Travel and Information Centers where travelers can get maps and brochures and make a rest stop. The Orange Visitors Center has the marsh as its focal point, and there is a boardwalk where you can stretch your legs after a long drive and have the opportunity to view wetland species, including alligators. The Visitors Center is located west of the Sabine River bridge on I-10 and is part of Tony Houseman State Park at Blue Elbow Swamp.

Sea Rim State Park Located in Jefferson County near the J. D. Murphree WMA, Sea Rim has over 4,000 acres of marshes. The Gambusia Nature Trail has a 3/4-mile (1.2 km) wooden boardwalk that goes through the marshes and offers a chance for viewing alligators and other wildlife up close. Airboat marsh tours are also available at the park that allow up to four people at a time to get a view of the marsh interior and its associated wildlife.

Federal Areas with Alligator Populations

Anahuac National Wildlife Refuge Anahuac consists of over 30,000 acres and is located off I-10. During the prime basking season, visitors may see a dozen or so alligators at a time. There are opportunities for alligator hunting on the refuge as well.

Aransas National Wildlife Refuge Aransas is located on the Gulf Coast, southeast of Victoria and southwest of Port Lavaca. Although home to alligators, the 59,000-acre refuge is best known for its endangered whooping cranes and nesting Kemp's ridley sea turtles. Matagorda Island, a barrier island, is part of this refuge.

McFaddin and Texas Point National Wildlife Refuges These refuges are located in Southeast Texas on the Gulf Coast along Texas 87 and include 50,000 acres of coastal marsh. They also feature several small lakes and bayous. They contain one of the densest populations of alligators in the state. McFaddin is located west of Sabine Pass along Highway 87. Texas Point is located at Sabine Pass and used to be known as Sea Rim. There are scheduled alligator hunts.

San Bernard National Wildlife Refuge Located on 34,679 acres in both Brazoria and Matagorda Counties, most of the refuge is not accessible to visitors. However, roads and hiking trails provide plenty of opportunities to view a rich variety of wildlife species, including alligators. The Pond Loop is the route where alligators may be viewed by vehicle.

Other Opportunities

Cattail Marsh in Tyrrell Park This wetland was constructed in 1993 by the City of Beaumont as the final phase of a wastewater treatment system. It is an example of how wetlands and the green movement can work together. It attracts over 350 bird species and other wildlife associated with wetlands, including alligators. There are group tours of the marsh in an open-air bus, which take place on specific days and require advance reservations.

Lake Jackson Wilderness Park This is a nearly 500-acre municipal park bordered by Buffalo Camp Bayou and the Brazos River. Alligators and other wildlife can be viewed at the park in their natural state. There is a 2-mile (3.2 km) hiking trail and boat access.

Captive Alligator Viewing

Texas has at present 16 AZA (American Zoo and Aquarium Association) facilities, and most of them display alligators. There are also a number of viewing opportunities in captivity that are not AZA institutions or even zoos. Some of these facilities can get a bit pricey for an entire family, so I have included some details on the background of each zoo and what it has to offer in addition to alligators. Some of these zoos have a history that involves alligators or that is of interest in general, so I have included some historical perspective as well.

Zoos with Alligator Displays:
AZA Institutions

Abilene Zoo The Abilene Zoo began in 1919 when the original zoo was built in Fair Park (now named Oscar Rose Park). The land that became the park was purchased for the first West Texas Fair and became Abilene's first municipal park. The zoo is still a division of the City of Abilene, which supplies its annual operations budget. The Abilene Zoological Society and the Grover Nelson Zoological Foundation are the sources of supporting donations. It has grown to include over 500 animals belonging to 200 + species. Alligators are displayed in the Wetlands of the Americas exhibit.

Caldwell Zoo The Caldwell Zoo at present consists of 85 acres in the Piney Woods of East Texas. It features multispecies exhibits of North and South America and Africa. The North American exhibit features various habitats, including an alligator pond. The zoo is home to over 2,000 animals representing some 250 species.

Cameron Park Zoo The Cameron Park Zoo is set up as a natural habitat zoo. For example, you can see zebra, giraffes, and antelopes running together as they would in the wild of the African savanna. In 2000, voters in Waco and McLennan Counties passed a $9.5 million bond issue to double the size of the zoo. Most of the bond money was spent to develop the new Brazos River Country exhibit, the signature exhibit for the zoo. Visitors encounter exhibits that represent specific habitats and some of the species associated with each. The swamp exhibit displays alligators, and they are visible underwater from a replica of a beaver lodge. Brazos River Country also includes a 50,000-gallon coral reef aquarium, coastal shorebirds, bears, coyotes, mountain lions, otters, a 35,000-gallon freshwater aquarium, a smaller aquarium with prehistoric-looking paddlefish, a butterfly garden, jaguars, ocelots, bats, owls, bison, deer, turkey, and javelina. Besides viewing the animals, kids can have additional fun experiences such as sliding down an acrylic tube through the otter exhibit and walking behind a waterfall.

Dallas World Aquarium The Dallas World Aquarium is housed in an unlikely location, an old warehouse built in 1924 near the historic West End District in downtown Dallas. The aquarium was opened to the public in 1992. In 1996, the adjacent building was purchased, which had been built in 1929 as a warehouse but had various uses throughout the years. Both buildings had been totally gutted, leaving the brick walls and supporting structure. The alley between the two buildings assumed the role of a "channel" between the freshwater and saltwater ecosystems. The aquarium contains 85,000 gallons of salt water and includes a 22,000-gallon walk-through tunnel. The exhibit Orinoco—Secrets of the River opened to the public in 1997. It is a rainforest exhibit featuring a number of bird species, mammals ranging from vampire bats to monkeys to manatees, amphibians, reptiles, invertebrates, and fish and includes a 40-foot waterfall. The highly endangered and rarely exhibited Orinoco crocodiles are a definite "must-see" in the exhibit. The vacant lot behind the original building was purchased in 2000 for the Mundo Maya exhibit, which opened in 2004 and features flora and fauna of importance to the Maya culture, including venomous snakes, sharks, jaguars, sea turtles, hummingbirds, owls, eagles, and fairy penguins. There is an outdoor South Africa exhibit and a seasonal Madagascar display. No alligators are displayed here, but it is definitely worth a visit to view one of the rarest crocodilians in the world.

Dallas Zoo The Dallas Zoo can boast of being the first Texas zoo, as well as the largest. Started in 1888, it has 95 developed acres of land. A new Ghosts from the Bayou alligator swamp exhibit opened April 18, 2009, in the Pierre A. Fontaine Bird & Reptile Building. A 9-foot (2.7 m) albino alligator named Boudreaux is displayed with two normally pigmented female alligators and with smaller exhibits of reptilian swamp residents: snapping turtles, copperheads, cottonmouth, and ratsnakes. There are two smaller albino alligators at the entrance to the exhibit, which features a tin-roofed shack with water trickling into a pond 3 feet (0.9 m) deep. The alligators have a 15-foot (4.5 m) pool surrounded by a sandy beach and muddy bank with cypress roots.

El Paso Zoo The 18-acre El Paso Zoo expanded to 36 acres with the addition of the new Reptile House and Passport to Africa exhibit. There are now some 1,700 animals that reside at the zoo. Alligators are featured in the Animals of the Americas exhibit.

Ellen Trout Zoo The Ellen Trout Lake has free-roaming alligators that bask on the shore seasonally. Fishing is permitted with a valid fishing permit. The zoo and lake are public facilities owned by the City of Lufkin. The zoo has nearly 700 reptiles, birds, and mammals and is currently undergoing its largest expansion phase since its initial opening.

Fort Worth Zoo The Fort Worth Zoo was founded

in 1909 with one lion, two bear cubs, an alligator, a coyote, a peacock, and a few rabbits. It was owned and operated by the City of Forth Worth until October 1991. Since then, the Fort Worth Zoological Association has taken over day-to-day operations. It has been named one of the top five zoos in America by *Family Life* magazine. It now houses over 5,000 animals. The American alligators at the Fort Worth Zoo are housed in the Texas Wild! Swamp exhibit with a large alligator snapping turtle.

Gladys Porter Zoo The Gladys Porter Zoo is located on 31 acres in central Brownsville. It was opened to the public in 1971 as a municipal zoo. However, it was not built by the city. It was planned, built, and stocked by the Earl C. Sams Foundation and then given to the city. Sams was a partner of J.C. Penney and was president of the J.C. Penney Company from 1917 to 1947. His daughter, Gladys, was president of the board of directors of the zoo until her death in 1980 and had been personally involved in the day-to-day operations of the zoo. The Herpetarium opened in 1973. The Herpetarium has a "pit" at its entrance with two American alligators. The zoo also has highly endangered Cuban and Philippine crocodiles, as well as saltwater crocodiles. Approximately 1,600 animals reside at the zoo. There are also a lot of plants throughout the zoo, which is especially known for its cactus garden.

Houston Zoo The Houston Zoo consists of 55 acres inside the 610 Loop. It became privatized in 2002. It is now operated by Houston Zoo, Inc., a not-for-profit organization. It currently houses more than 3,500 exotic animals representing more than 700 species. The cur-

rent alligator exhibit is located in the Reptile House and showcases a leucistic alligator. The Children's Zoo for many years used several juvenile alligators for education programs on the zoo grounds and off-site demonstrations with docents on Zoomobile trips to places such as hospitals.

Landry's Downtown Aquarium—Houston An AZA-accredited zoo run by a restaurant corporation? Unlikely but true. Landry's Restaurants Inc., based in Houston, owns the Downtown Aquarium in Houston, as well as the Downtown Aquarium—Denver and an aquarium restaurant in Nashville. The 20-acre Downtown Aquarium opened in 2003, an entertainment complex with a public aquarium, two restaurants, a bar, banquet facilities, amusement rides, and midway games. The aquarium includes a 100,000-gallon, floor-to-ceiling centerpiece aquarium, which is the tallest cylindrical tank in North America; a 200,000-gallon shark tank; white tigers displayed in a temple ruins exhibit; a rain forest with a variety of vertebrate species; a shark nursery; and a swamp exhibit that includes American alligators. There is also a chance for hands-on experiences with stingrays and invertebrates, such as horseshoe crabs. In addition to dinner, a typical amount of time spent seeing all that the Downtown Aquarium has to offer is two to three hours. Expect long waits at the restaurants and use that time to see the exhibits.

San Antonio Zoological Gardens & Aquarium In 1914, Colonel George Brackenridge, a prominent citizen who founded the *San Antonio Express-News*, assembled buffalo, elk, deer, monkeys, a pair of lions, and four bears on land that he deeded to the city. This land became known as Brackenridge Park, and the collection of animals became the San Antonio Zoo. The zoo is located at the headwaters of the San Antonio River and has artesian water flowing through the zoo grounds. The site is a former rock quarry, which provides limestone cliffs for naturalistic exhibit settings. The humid, tropical conditions along the riverbank, in addition to the semi-arid hillsides, create a variety of habitats. This factor, along with the city's temperate climate, makes it possible to keep most animals outside year-round. The zoo has over 3,500 animals consisting of 600 + species on 56 acres. There are currently three alligators on exhibit.

SeaWorld San Antonio The world's largest marine life park, occupying 250 acres, opened in 1988. It was originally known as SeaWorld of Texas, and its name was changed when Anheuser-Busch Inc. took it over. Of course, Anheuser-Busch Clydesdale draft horses took up residence at the park as well. Other non-marine animals include a variety of birds and American alligators. The alligators can be found in an exhibit known as Alligator Alley, located near Ski Lake. There are approximately 20 alligators, some turtles, and fish.

Texas State Aquarium The Texas State Aquarium in Corpus Christi opened to the public in 1990. In 1999, a special exhibit known as Swamp Tales opened, featuring Roscoe the white alligator. At present, a new Swamp Tales exhibit is under construction. The facility currently features an 11-foot (3.3 m) American alligator, and there are opportunities for the public to meet smaller alligators.

Present Displays: Non-AZA Institutions

Animal World and Snake Farm The Snake Farm (now known as the Animal World and Snake Farm) in New Braunfels has been in existence since 1967. Ownership changed in the 1990s, and has recently changed again. There are over 500 resident animals: primates, alligators, exotic hoofstock, birds, cats, and snakes. One popular feature is an alligator lagoon located near the picnic area. In the summer, crocodile and alligator shows and feedings are offered.

Bayou Wildlife Park Clint Wolston has been developing his private animal reserve in the Houston suburb of Alvin for 30 years, and it is now an outstanding park populated by free-roaming birds and animals from around the world. Wolston says that over this time he has learned "things" not found in many books. There are trams to take visitors through 86 acres of prairie and woods. See more than 500 animals representing 50 species of mammals and birds. The park includes alligator ponds.

The Crocodile Experience This facility in Angleton features a boardwalk that allows visitors to get up close to Nile crocodiles. The visitor can see and interact with American alligators and other crocodilian species, as well as other reptiles. Visitors are allowed to enter a pen with large tortoises to pet and feed. There are even a few non-reptiles, including eland (the largest antelope in the world). Owner Chris Dieter is a high school science teacher, and he has made his facility both educational and fun for all ages. Animal displays can also be brought to your school or event.

Gator Country Wildlife Adventure Park and Restaurant Gator Country in Beaumont is the site of a former alligator farm called Alligator Island that was billed as the largest breeding farm in Texas and featured a large alligator named Big Al. Gary Saurage and his partners purchased the farm in 2005 and renovated it. It reopened as an alligator theme park with a restaurant that is in close proximity to one of the three alligator ponds in the park. The park is home to nearly 300 alligators, several other crocodilian species, and others. There are educational shows, a chance to touch baby alligators, and great alligator viewing.

Texas Alligator Ranch Dean Coates operates this not-for-profit educational facility that features more than 24 alligators. The Ranch is located a quarter-mile west of Spearman on Highway 15.

The Texas Zoo The Texas Zoo was voted "The National Zoo of Texas" in 1984 by the Texas legislature. The zoo got its start in 1957 with a lioness, whose cage was donated by the Lions Club. It was first known as the Victoria City Zoo, and its name was changed in 1976. A naturalistic swamp exhibit features alligators, and visitors enter a viewing area and look through a window to see them. During the warmer months, the alligators are fed on Saturdays at 1:00 p.m., and the visitors are invited to watch.

Non-Zoos

First National Bank of Alvin An alligator exhibit in a bank? Alvin is home to several alligators and a snapping turtle. It all started years ago when a customer put a hatchling alligator in the fishpond in the main lobby of the bank. The bank ended up getting two more and then obtained a permit to keep them. A snapping turtle later joined them. This glassed enclosure called the

Figure 7.1. The First National Bank of Alvin has a central courtyard with a pool that is home to alligators. The bank is probably the most unusual site for a reptile exhibit.

Alligatrium served as their home until 1980, when the bank moved into a new building with a more spacious Alligatrium (fig. 7.1). The best vantage point is to take the elevator upstairs and look down on the entire exhibit. The bank uses the gator concept in their logo and marketing. For example, they coined the phrases "tel-a-gator" and "gator banking."

Gatorfest Think of a small-town festival with food, rides, and craft booths, plus some entertainment and pageants. Gatorfest is held in Anahuac, which was designated the "Alligator Capital of Texas" by the Texas legislature, as the reptilian residents outnumber the humans almost 3 to 1. The festival site is Fort Anahuac State Park, located at the mouth of the Trinity River. Gatorfest takes place each September on the opening weekend (Friday–Sunday) of the alligator hunt. This coincides with the Great Texas Alligator Roundup, which allows Texas alligator hunters to bring their alligators in to compete for cash prizes. It also gives them a way to sell their alligators if they choose to do so, as a licensed Texas alligator buyer is on-site. TPWD has an educational booth with young alligators that the public can touch. There are also opportunities for airboat rides. Vendors offer alligator products and food booths where visitors have the opportunity to eat alligator prepared various ways.

Displays of the Past

Cement Alligator Pond, Sociable Inn Tavern, Elmendorf This exhibit in Bexar County was not one that any decent, law-abiding citizen would have fond memories of. The saloon was owned by a man named Joe Ball, who was documented to have murdered at least two women and was reputed to have been a serial killer. Furthermore, he was alleged to have fed the bodies of his victims to the alligators. The movies by

Tobe Hooper, *Texas Chainsaw Massacre* and *Eaten Alive*, are believed to be based on Joe Ball's story. The horror stories concerning this man were handed down through generations and told as ghost stories around campfires. Fact was difficult to discern from fiction, and undoubtedly details were embellished. The true story was difficult to come by, as there were no records or published accounts in existence from the time that the story took place. The individuals involved in the investigation died long ago.

We now know a lot more about "The Butcher of Elmendorf" thanks to Michael Hall, managing editor of the *Austin Chronicle*, who did the research and published the story in the July 1, 2002, issue of *Texas Monthly*. His research was based largely on interviews with Ball's relatives and witnesses who were still alive. Joe Ball was a loner as a child and spent time fishing and exploring. He became extremely interested in guns and was an excellent shot. He enlisted in the military and participated in World War I. He received an honorable discharge and returned home to work in his father's general store.

He left his father's store and worked as a bootlegger during Prohibition. After Prohibition ended, Joe opened a tavern named the Sociable Inn. He had a cement alligator pond put in back of the saloon and placed five alligators in it. The alligators drew a lot of interest, and on Saturdays, a drunken crowd could watch the alligators being fed.

Stories began circulating about barmaids and others disappearing. It was rumored that the disappearing individuals were being fed to the alligators. At one point, Ball was involved with three barmaids and married one of them. All three ended up missing. The woman he married turned up alive in California. The other two were confirmed victims, whose remains were recovered. Before the bodies were discovered, Ball was to be taken in for questioning on September 24, 1938. He asked if he could close up his bar first. The investigators complied. Ball went over to his cash register, took out a revolver, and committed suicide.

The investigators never found evidence that Ball fed any victims to the alligators. The alligators went to live at the San Antonio Zoo. Another missing barmaid turned up alive in Arizona. Did Joe Ball kill more than two people? Probably.

Houston Zoo In 1905, the population of Houston numbered 44,000 people and 80 automobiles. C. E. Barrett had just made an oil strike in the town of Humble, located just northeast of Houston, and this event began Houston's entrance into the energy business. Sam Houston Park, situated at the edge of downtown, not far from Buffalo Bayou, was the location of the city's first zoo. The collection consisted of rabbits, raccoons, Mexican eagles, black bear, a great horned owl, capuchin monkeys, prairie dogs, and an alligator pond. In 1920, the federal government donated a bison named Earl to the City of Houston, and he was added to the zoo's growing number of animals at Sam Houston Park.

Earl's arrival sparked a debate in the city that the zoo should be improved and expanded into a world-class zoo. The following year the city purchased snakes, birds, and alligators and in 1922 fenced a tract of land in Hermann Park

to start the new zoo. Hans Nagel was hired as the first zookeeper to care for the 40 animals. By 1925, Nagel became the director of the zoo, the zoo's acreage increased, and the animal population grew to 400. Over the years, the animals, personnel to care for them, and exhibits increased in number. With the arrival of zoo director John Werler, education was emphasized, and a docent program with over 100 volunteers was organized in the 1970s.

The sea lion pond, built in 1925 to house three donated sea lions, was eventually turned into an alligator exhibit. It consisted of a cement pool, surrounded by grassy areas that allowed alligators to come out of the water and bask close to the public. A chain-link fence kept the alligators in the enclosure. Public feeding sessions became popular with zoo visitors. In the early 1990s, a new swamp exhibit for the alligators was built on the site, which was more naturalistic. The exhibit was removed several years ago to make way for a new elephant habitat.

Many Stones Rock Shop, Terlingua These are the words of Ring Huggins, who warned me that he was a storyteller (October 21, 2004):

Five years ago this month I moved back to far West Texas from Canada after 30 years of running a craft shop, working construction, numerous very odd jobs, and many winters teaching adult education on remote Native reserves (what Gringos call Indian reservations) in British Columbia and what was then called the Northwest Territories. I rented what was once the old Chevron Gas Station at Study Butte, which belonged to Warren "Gator" Lynch to use as a rock and craft store. The only condition aside from normal commercial rental agreements was that he expected me to care for his blind six-foot-long female alligator. How does an alligator wind up in the far West Texas desert, you may ask. More on that later.

Warren had moved down on the Rio Grande south of the Black Gap and the Stilwell Ranch to run a small resort with his wife, Roberta, for a retired mining engineer, Andy Curry. Miss Tracy, who had rented the café next door from them, was caring for the alligator when Warren was elsewhere. For me, caring for the animal was no problem since one of my very odd jobs was working as a zookeeper in the B.C. Wildlife Park in Kamloops, and I cared for numerous reptiles, including caimans. My problem with the female alligator became twofold. Firstly, I wound up spending about two hours per day talking to tourists and locals about alligators. I am not in the alligator business. I am in the rock and craft business. Secondly, it was costing me a lot of money to pump the well to keep the alligator pond full, and I got no rent credits for my efforts and expenses.

After renting the property for several months, caring for the gator, and having made numerous improvements to the place, I thought that I should make arrangements to buy the place, having had several rental spaces I improved at great expense sold out from under me. During this time, Roberta got mad at "Gator"

(Warren) and had him thrown in the Ector County Holding Facility in Odessa for spousal abuse. We are not sure if Warren ever did this abuse, but he was a felon and the Ector County sheriff arrested him. During his incarceration, I corresponded with him, writing numerous letters. Warren's felony concerned alligators.

I wrote to Warren and Roberta, making them a cash offer they did not refuse, and also stated that I had been researching alligator meat recipes on the Internet. Warren was here in less than a week, getting my drift, and took the reptile elsewhere. A few weeks later I owned the property, the alligator was elsewhere, and all was well with the world.

Reptile Garden at the Witte Museum, San Antonio A detailed history of the reptile garden is included in Bess Woolford and Ellen Quillin's book *The Story of the Witte Museum* (1966). The Reptile Garden came about during the Great Depression as the result of a Welsh herpetologist named W. C. Bevan, who wanted a permanent job at the museum. He suggested building a reptile garden, and the director and museum board (all women) went for the idea because they had nothing to lose. They had only $2.47 left to pay maintenance costs for the next three months. The original garden consisted of a wooden fence surrounded by barbed wire and was located at the side of the museum near an irrigation ditch. Discarded metal roofs were used to shade the snake pit, and a pool was built. A bridge was constructed for visitors to view the reptiles. Although there was a variety of reptiles put in the exhibit, rattlesnakes were the mainstay. Out-of-work individuals and ranchers from the Southwest and Mexico brought rattlesnakes to sell by the pound. The exhibit was popular from the beginning and paid for itself during the first week (it cost $750 to build).

A new Reptile Garden was built out of rock with a red tile roof by 1934. The exhibit saved the museum from financial ruin during the Depression and also became a center for herpetology. Various snake handlers gave shows, and it provided a source of snakes for researchers to use in developing techniques for treating snakebite. There were also turtle races, where the public could pay to bet on a turtle of their choice out of a field of about 30.

Although alligators were on exhibit from the start, many were brought in during a drought year, and about 80 were present at their peak. One even nested. There is a great photograph in Woolford and Quillin's book of alligators with rattlesnakes climbing over their backs. In 1950, the Reptile Garden closed. The snakes had become hard to obtain since people no longer were catching snakes for a living, the San Antonio Zoo had a reptile house, and the museum's finances improved. The reptiles, except for the alligators, were given to the zoo. The alligators and the exhibit were turned over to a private firm to run and reopened as the Alligator Garden in 1952 (fig. 7.2). It was managed by George Kimbrell, who collected and displayed alligators and crocodiles.

Figure 7.2. The Witte Museum reptile exhibit, noted for its snakes and alligators, became the Alligator Garden in 1952. This vintage postcard shows the entrance to the facility.

Lonnie McCaskill, zoological manager at Disney's Animal Kingdom, has childhood memories of the Alligator Garden. He went to the garden when he visited his grandparents and credits those visits as one of the reasons that he chose to work with crocodilians and other reptiles. The man who ran the Alligator Garden took him on tours whenever he visited and sometimes gave him shed teeth. One alligator had a spinal deformity, and people put coins in its pond. The man said that he donated the money "to the little kids' polio." Lonnie thought that the owner said that he was sending the alligators to the Oklahoma Zoo when he closed, but he wasn't able to confirm that. An account that I read said that the Alligator Garden closed in 1975 when Kimbrell retired to Arkansas and that he took his collection with him.

San Jacinto Plaza or La Plaza de los Lagartos, El Paso The plaza was originally the site of the corrals for Ponce de Leon's ranch. It was donated to El Paso in 1881 by William T. Smith, who had bought the land from Ponce de Leon's heirs. The City contracted with El Paso Parks and Streets Commissioner J. Fisher Satterthwaite to create a beautiful plaza, and by 1883 Satterthwaite created a park surrounded by a fence, containing a walled pond, a gazebo, and a planting of 75 Chinese elm trees. It was named San Jacinto Plaza by the city council in 1903.

Some accounts say that Satterthwaite brought three alligators to El Paso and kept them in a barrel of water until a pond could be built in the plaza (fig. 7.3). Another story is that in about 1883 a miner named Adolph Munsenberger received a shoebox-sized package from a friend in Louisiana. It contained six baby alligators along with a note that said to use the alligators to haul mining carts. He put them in a water barrel to raise them but knew that they would soon become large. He offered them to the City to be put in the newly completed pond in San Jacinto Plaza.

Some accounts state that during the cold winter months when the alligators were young, they were wrapped in coffee sacks and taken to the saloon and placed near stoves to keep them warm. There are also records that they would retreat to the center of the pond in late October and remain in the water until March. It was viewed as a sign of spring when they came out in the open again. From various accounts, it seems that at least three alligators were placed in the plaza pond and that seven was the maximum number at one time.

The alligators became local celebrities and had names and personalities. There was even a weekly column in local newspaper called "Alligators' Tails." Unfortunately, they were "out there" for people to abuse as well. An alligator from the plaza was placed in a hotel room to "welcome" its guest. On another occasion, geology professor Howard Quinn at Texas Western College had one of the alligators placed inside his office. I met Jeannie Murray at a Red Hat Society meeting, and she told me that during her time as a student at UTEP, one morning her dorm awakened to a commotion of police and fire trucks. They were called to the university by pool attendants to remove an alligator that apparently a few boys had kidnapped from the plaza and placed in the swimming pool.

Figure 7.3.
(a, b) Vintage
postcards of the
El Paso Plaza
showing the
alligator exhibit.

After a number of attacks by vandals, the alligators were moved to the El Paso Zoo for their own safety. In 1967, the pond was torn down, and the alligators may have been removed as early as 1965. Some accounts say that the alligators made a brief return from 1972 to 1974, protected by a plastic enclosure, only to be returned to the zoo again due to vandals.

Luis Jimenez, an artist who grew up in El Paso and remembered the alligators at the plaza from his childhood, was commissioned to create a fiberglass sculpture. *La Plaza de los Lagartos* features alligators lunging outward as part of a fountain. This is the centerpiece of the remodeled plaza (fig. 7.4). It is also a favorite spot for birds to come and take a drink and rest a while.

Sea-Arama Marineworld, Galveston Sea-Arama opened in 1965 as an ocean theme park. It was over 25 acres in size, had a 4-acre ski lake, and a 50-foot-long, 200,000-gallon aquarium. There were ski shows, shark feeding shows, marine mammal shows, bird shows, and even an alligator wrestling show. In 1988, it was the number-one tourist attraction in Galveston. Unfortunately, the opening of the larger Sea-World in San Antonio outcompeted it, and the park closed in 1990 and was demolished. Photographs, videos, and memories of Sea-Arama have been posted at http://galveston.wix.com/rememberingseaarama#!about1/copg.

Figure 7.4. (a) El Paso Plaza showing the alligator exhibit, probably in the 1950s. (Photo provided by Jan Campbell and Eugenia Murray). (b) The Plaza as it looks today with Luis Jimenez's sculpture. (Photo by Lisa McDonald)

a

b

References

General References

Bakker, R. T. 1986. The Dinosaur Heresies. New York: Kensington Publishing.

Chatfield, J. 1995. A Look inside Reptiles. Pleasantville, NY: Reader's Digest Young Families.

Chiappe, L. M., and L. Dingus. 2001. Walking on Eggs: The Astonishing Discovery of Thousands of Dinosaur Eggs in the Badlands of Patagonia. New York: Scribner.

Chiasson, R. B. 1962. Laboratory Anatomy of the Alligator. Dubuque, IA: William C. Brown.

Guggisberg, C. A. W. 1972. Crocodiles: Their Natural History, Folklore and Conservation. Harrisburg, PA: Stackpole Books.

Johnson, L. A., A. Cooper, B. Thompson, and R. Wickwire. 1989. Texas Alligator Survey, Harvest, and Nuisance Summary 1988. In Crocodilian Congress Production and Marketing, 36–72.

Lillywhite, H. B. 2008. Dictionary of Herpetology. Malabar, FL: Krieger Publishing.

Onadeko, S. A. 1983. Status of the American Alligator and Potential Resource Management Problems at Brazos Bend State Park. Master's thesis, Texas A&M University.

Ouchley, K. 2013. American Alligator: Ancient Predator in a Modern World. Gainesville: University Press of Florida.

Potter, F. E., Jr. 1974. Population Status of the American Alligator in Texas. Special Report, Texas Parks and Wildlife Department. Mimeo.

———. 1975. American Alligator Study. Special Report, Texas Parks and Wildlife Department.

———. 1981. Status of the American Alligator in Texas. Special Report, Texas Parks and Wildlife Department.

Ross, C. A., ed. 1989. Crocodiles and Alligators. New York: Facts on File.

Texas Parks and Wildlife Department. 1987. Texas Alligator Survey, Harvest, and Nuisance Summary. Available at https://repositories.tdl.org/tamug-ir/handle/1969.3/21914.

References Cited

Allen, E. R., and W. T. Neill. 1956. Some Color Abnormalities in Crocodilians. Copeia 1956 (2): 124.

Alvarez del Toro, M. 1969. Breeding the Spectacled Caiman, Caiman crocodilus, at Tuxtla Gutierrez Zoo. Int. Zoo Yearbook 9:35–36.

Anon. 1962. Pet Caiman—Relationship with Cat. Animal Kingdom, February, 26–27.

Anon. 1901. The Reminiscences of Mrs. Dilue Harris. Southwest. Hist. Quart. 4 (3): 155–189.

AP Press. 2003. Gator Blindness Increase Seen; Scientists Unsure Why. Clarion-Ledger (Jackson, MS), July 1.

Behler, J. L., and F. W. King. 1979. The Audubon Society Field Guide to North American Rep-

REFERENCES

▼▼▼▼▼

208

tiles and Amphibians. New York: Alfred A. Knopf.

Berlandier, J. L. 1980. Journey to Mexico during the Years 1926 to 1834. 2 vols. Austin: Texas State Historical Association and University of Texas at Austin.

Bossu, B. n.d. New Voyages in North America, 1751–62. W.P.A. translation in typescript. New Orleans: Louisiana Historical Center, Louisiana State Museum.

———. 1982. New Travels in North America, 1751–62. Trans. S. D. Dickinson. Natchitoches: Northwestern State University of Louisiana Press.

Bradshaw, J. 2013. Cat Sense: How the New Feline Science Can Make You a Better Friend. New York: Basic Books.

Brantley, C. G. 1989. Food Habits of Juvenile and Sub-adult American Alligators (*Alligator mississippiensis*) in Southeastern Louisiana. Master's thesis, Southeastern Louisiana University.

Broussard, J. C. 1988. Efficiency of Collecting, Incubating, and Hatching Eggs from Wild Alligator Nests in Texas as a Means of Supply for Alligator Farmer Stock. Consultant Services Agreement Contract Report (included with the TPWD Texas Alligator Annual Report for 1988). Austin: Texas Parks and Wildlife Department.

Chabreck, R. H. 1971. The Foods and Feeding Habits of Alligators from Fresh and Saline Environments in Louisiana. Proc. Ann. Conf. Southeast. Assoc. Fish Wildlife Agencies 25:117–124.

Colbert, E. H., R. B. Cowles, and C. M. Bogert. 1946. Temperature Tolerances in the American Alligator and Their Bearing on the Habits, Evolution, and Extinction of the Dinosaurs. Am. Mus. Nat. Hist. 86:329–373.

Conant, R., and J. T. Collins. 1998. The Peterson Field Guide to Reptiles and Amphibians of Eastern and Central North America. 3rd ed. Boston: Houghton Mifflin.

Cott, H. B. 1961. Scientific Results of an Inquiry into the Ecology and Economic Status of the Nile Crocodile in Uganda and N. Rhodesia. Trans. Zool. Soc. Lond. 29:211–356.

Coulson, R. A., J. D. Herbert, and T. D. Coulson. 1989. Biochemistry and Physiology of Alligator Metabolism *in Vivo*. Biology of the Crocodilia. Am. Zool. 29 (3): 921–934.

Coulson, R. A., and T. Hernandez. 1964. Biochemistry of the Alligator. Baton Rouge: Louisiana State University Press.

Cowardin, L. M., V. Carter, F. C. Golet, and E. T. LaRoe. 1979. Classification of Wetlands and Deepwater Habitats of the United States. Washington, DC: USDI Fish and Wildlife Service.

Crain, J. K. 1957. A Texas Family. Available at Sons of DeWitt Colony Texas, http://www.tamu.edu/faculty/ccbn/dewitt/dewitt.htm.

Davis, L. M., T. C. Glenn, R. M. Elsey, H. C. Dessauer, and R. H. Sawyer. 2001. Multiple Paternity and Mating Patterns in the American Alligator, *Alligator mississippiensis*. Mol. Ecol. 10:1011–1024.

Deeming, D. C., and M. J. W. Ferguson. 1989. The Mechanism of Temperature-Dependent Sex Determination in Crocodilians: A Hypothesis. Am. Zool. 29 (3): 973–985.

Delany, M. F. 1990. Late Summer Diet of Juvenile American Alligators. I. Herpetology 24 (4): 418–421.

Delany, M. F. and C. L. Ambercrombie. 1986.

American Alligator Food Habits in North Central Florida. J. Wildlife Manage. 50:348–353.

Dinets, V. J. C., and J. D. Brueggen. 2015. Crocodilians Use Tools for Hunting. Ethol. Ecol. Evol. 27 (1). Available at http://www.tandfonline.com/doi/abs/10.1080/03949370.2013.85827 6#.Va1BKPlmGNO.

Dixon, J. R., and M. A. Staton. 1976. Survivorship of Alligators through Their First Year and One Half of Life. Final Report, US Fish and Wildlife Service.

Elsey, R. M., and C. Aldrich. 2009. Long-Distance Displacement of a Juvenile Alligator by Hurricane Ike. Southwest. Nat. 8 (4): 746–749.

Erickson, G. M., A. K. Lappin, and K. A. Vliet. 2003. The Ontogeny of Bite-Force Performance in American Alligator (*Alligator mississippiensis*). J. Zool. Lond. 260:317–327.

Ferguson, M. W. J. 1985. Reproductive Biology and Embryology of the Crocodilians. In Biology of the Reptilia, vol. 14, Development A, ed. C. Gans, F. S. Billet, and P. F. A. Maderson, 331–491. New York: John Wiley and Sons.

Ferguson, M. W. J., and T. Joanen. 1982. Temperature of Egg Incubation Determines Sex in *Alligator mississippiensis*. Nature 296:850–853.

———. 1983. Temperature-Dependent Sex Determination in *Alligator mississippiensis*. J. Zool. Lond. 200:143–177.

Ferris, Timothy. 2003. Seeing in the Dark: How Backyard Stargazers Are Probing Deep Space and Guarding Earth from Interplanetary Peril. New York: Simon and Schuster.

Flack, Captain. 1866. The Alligator. In A Hunter's Experiences in the Southern States of America, chap. 26. London: Longmans, Green.

Foster, W. C. 1998. The LaSalle Expedition to Texas. In The Journal of Henri Joutel 1684–1687, 169. Austin: Texas State Historical Association.

Franklin, R. 1998. Alligator Farming: Texas. Houston Chronicle magazine, October 18, 8–13.

Garrick, L. D., and H. A. Herzog Jr. 1978. Social Signals of Adult American Alligators. Bull. Am. Mus. Nat. Hist. 160 (3): 153–192.

Garrick, L. D., and J. W. Lang. 1977. Social Signals and Behaviors of Adult Alligators and Crocodiles. Am. Zool. 17:225–239.

Glasgow, V. L. 1991. A Social History of the American Alligator. New York: St. Martin's Press.

Graham, A., and P. Beard. 1973. Eyelids of Morning: The Mingled Destinies of Crocodiles and Men. New York: A&W Visual Library.

Graham, F., and A. Graham. 1979. Alligators. New York: Delacorte Press.

Grant, C. 1956. Alligators in Western Texas. Herpetologica 12 (2): 90.

Griffin, S. C. 1931. History of Galveston, Texas: Narrative and Photographical. Galveston: H. Cawston.

Hagan, J. M., P. C. Smithson, and P. D. Doerr. 1983. Behavioral Response of the American Alligator to Freezing Weather. J. Herpetol. 17 (4): 402–404.

Hardin, S. L. 1991. A Hard Lot: Texas Women in the Runaway Scrape. East Tex. Hist. J. 29 (1): 35.

Hayes, L. 1992. Some Aspects of the Ecology and Population Dynamics of the American Alligator in Texas. PhD diss., Texas A&M University.

Hayes-Odum, L., L. A. Bailey, T. Hill-Kennedy, D. Cowman, and P. Reiff. 1994. *Alligator mississippiensis* (American Alligator) Nests. Herpetol. Rev. 25 (3): 119.

Hayes-Odum, L., and J. R. Dixon. 1997. Abnormalities in Embryos from a Wild American Alligator (*Alligator mississippiensis*) Nest. Herpetol. Rev. 28 (2): 73–75.

Hayes-Odum, L., T. Hill-Kennedy, L. A. Bailey, D. Cowman, and P. Reiff. 1996. *Alligator mississippiensis* (American Alligator) Reproduction. Herpetol. Rev. 27 (4): 199–200.

Hayes-Odum, L., and D. Jones. 1993. Effects of Drought on American Alligators (*Alligator mississippiensis*) in Texas. Tex. J. Sci. 45 (2): 182–185.

Hayes-Odum, L., D. Valdez, M. Lowe, L. Weiss, P. H. Reiff, and D. Jones. 1993. American Alligator (*Alligator mississippiensis*) Nesting at an Inland Texas Site. Tex. J. Sci. 45 (1): 51–61.

Huchzermeyer, F. W. 2003. Crocodiles: Biology, Husbandry and Diseases. Cambridge, MD: CABI Publishing.

Hunt, R. H. 1987. Nest Excavation and Neonate Transport in Wild *Alligator mississippiensis*. J. Herpetol. 21 (4): 348–350.

International Shark Attack File. Accessed July 20, 2015. Available at http://www.flmnh.ufl.edu/fish/sharks/statistics/statistics.htm.

Jacobson, E. R. 1989. Diseases of Crocodilians: A Review. In Crocodilian Congress Production and Marketing, 24–35.

Joanen, T., and L. McNease. 1979. Time of Egg Deposition for the American Alligator. Proc. Ann. Conf. Southeast. Assoc. Fish Wildlife Agencies 33:15–19.

———. 1989. Ecology and Physiology of Nesting and Early Development of the American Alligator. Biology of the Crocodilia. Am. Zool. 29 (3): 987–998.

Joanen, T., L. McNease, and G. Perry. 1977. Effects of Simulated Flooding on Alligator Eggs. Proc. Ann. Conf. Southeast. Assoc. Fish Wildlife Agencies 31:33–35.

Johnson, L. A., and B. C. Thompson. 1986. Evaluation of Alligator Hunter Capabilities to Retrieve Alligators Shot While Free-Swimming (Free-Ranging). In 1985 Texas Parks and Wildlife Department Annual Report, 3–5. Austin: Texas Parks and Wildlife Department.

Jones, D., and L. A. Hayes-Odum. 1990. *Alligator mississippiensis* (American Alligator) Behavior. Herpetol. Rev. 21 (3): 59–60.

———. 1994. A Method for the Restraint and Transport of Crocodilians. Herpetol. Rev. 25 (1): 14–15.

Jones, M. 1988. A Front Row View. The Spoonbill 37 (8): 5.

Kushlan, J. A., and M. S. Kushlan. 1980. Function of Nest Attendance in the American Alligator. Herpetologica 36:27–32.

Lang, J. K. 1987a. Crocodilian Behavior: Implications for Management. In Wildlife Management of Crocodiles and Alligators, ed. G. J. W. Webb, S. C. Manolis, and P. J. Whitehead, 273–294. Sydney, Australia: Surrey Beatty and Sons.

———. 1987b. Crocodilian Thermal Selection. In Wildlife Management of Crocodiles and Alligators, ed. G. J. W. Webb, S. C. Manolis, and P. J. Whitehead, 301–317. Sydney, Australia: Surrey Beatty and Sons.

Lee, D. 1968. Possible Communication between Eggs of the American Alligator. Herpetologica 24:88.

McIlhenny, E. A. 1935. The Alligator's Life History. Boston: Christopher Publishing House.

McNease, L. and T. Joanen. 1977. Alligator Diets in Relation to Marsh Salinity. Proc. Ann.

Conf. Southeast. Assoc. Fish Wildlife Agencies 31:36–40.

Mearns, E. A. 1907. Mammals of the Mexican Boundary of the United States. U.S. Nat. Mus. Bull. 56.

Minton, S. A., Jr., and M. R. Minton. 1973. Giant Reptiles. New York: Charles Scribner's Sons.

Mittleman, M. B., and B. C. Brown. 1948. The Alligator in Texas. Herpetologica 4 (6): 195–196.

Modha, M. L. 1967. The Ecology of the Nile Crocodile on Central Island, Lake Rudolf. East African Wildlife J. 5:74–95.

Neill, W. T. 1971. The Last of the Ruling Reptiles: Alligators, Crocodiles, and Their Kin. New York: Columbia University Press.

Ogden, J. C. 1971. Survival of the American Crocodile in Florida. Anim. Kingdom 74:7–11.

Ogden, J. C., and C. Singletary. 1973. Night of the Crocodile. Audubon 75:32–37.

Pike, D. A., and R. B. Smith. 2005. *Alligator mississippiensis* (American Alligator). Mortality. Herpetol. Rev. 36 (1): 61–62.

Platt, S. G., C. G. Brantley, and R. W. Hastings. 1990. Food Habits of Juvenile American Alligators in the Upper Lake Pontchartrain Estuary. Northeast Gulf Sci. 11 (2): 123–130.

Raun, G. G., and F. R. Gehlbach. 1972. Amphibians and Reptiles in Texas. Dallas Mus. Nat. Hist. Bull. 2.

Reese, A. M. 1915. The Alligator and Its Allies. New York: G. D. Putnam and Sons.

Reisner, M. 1991. Game Wars: The Undercover Pursuit of Wildlife Poachers. New York: Viking Penguin.

Saumure, R. A., B. Freiermuth, J. Jundt, L. Rowlett, and J. Jewell. 2002. A New Technique for the Safe Capture and Transport of Crocodilians in Captivity. Herpetol. Rev. 33 (4): 294–329.

Scott, T. P., S. R. Simcik, and T. M. Craig. 1999. A Key to Some Pentastome, Nematode and Trematode Parasites of the American Alligator. Tex. J. Sci. 51 (2): 127–138.

Scott, T. P., and P. J. Weldon. 1990. Chemoreception in the Feeding Behavior of Adult American Alligators, *Alligator mississippiensis*. Anim. Behav. 39 (2): 398–400.

Smith, H. M., and D. Chiszar. 2003. Observations by Berlandier 1827–1834 on the Crocodilians of Texas and Northeastern Mexico. Bull. Chic. Herpetol. Soc. 38 (8): 155–157.

Snider, A. T., and J. K. Bowler. 1992. Longevity of Reptiles and Amphibians in North American Collections. 2nd ed. SSAR Herpetol. Circ. 21.

Soares, D. 2002. An Ancient Sensory Organ in Crocodilians. Nature 417:241–242.

Somma, L. A. 2003. Parental Behavior in Lepidosaurian and Testudinian Reptiles. Malabar, FL: Krieger.

Stejneger, L., and T. Barbour. 1943. A Check List of North American Amphibians and Reptiles. 5th ed. Bull. Mus. Comp. Zool., Harvard Univ. 93:1–260.

Strecker, J. K. 1915. Reptiles and Amphibians of Texas. Baylor Bull. 18 (4): 1–82.

Sullivan, B. 1994. Scales of Justice Tip in Alligator's Favor. Chicago Tribune, March 13. Available at http://articles.chicagotribune.com/1994-03-13/features/9403130264_1_alice-swede-pet-store.

Taylor, D. n.d. An Alligator Population Model and Associated Minimum Population Estimate for Non-marsh Alligator Habitat in Louisiana. Unpublished report. Louisiana Department of Wildlife and Fish.

REFERENCES

▼▼▼▼▼

212

Taylor, D., and W. Neal. 1984. Management Implications of Size-Class Frequency Distributions in Louisiana Alligator Populations. Wildlife Soc. Bull. 12 (3): 312–319.

Taylor, J. A., G. J. W. Webb, and W. E. Magnusson. 1978. Methods of Obtaining Stomach Contents from Live Crocodilians (Reptilia, Crocodilidae). J. Herpetol. 12 (3): 415–417.

Thompson, B. C., F. E. Potter Jr., and W. C. Brownlee. 1984. Management Plan for the American Alligator in Texas. Austin: Texas Parks and Wildlife Department.

Thompson, R. L., and C. S. Gidden. 1972. Territorial Basking Counts to Estimate Alligator Populations. J. Wildlife Manage. 36 (4): 1081–1088.

Valenzuela, N., and V. A. Lance, eds. 2004. Temperature-Dependent Sex Determination in Vertebrates. Washington, DC: Smithsonian Institution Press.

Viosca, P., Jr. 1939. External Sexual Differences in the Alligator, *Alligator mississippiensis*. Herpetologica 1:154–155.

Vliet, K. A. 1989. Social Displays of the American Alligator (*Alligator mississippiensis*). Am. Zool. 29 (3): 1019–1031.

Webb, G. J. W., S. C. Manolis, and P. J. Whitehead, eds. 1987. Wildlife Management of Crocodiles and Alligators. Sydney, Australia: Surrey Beatty and Sons.

Webb, G. J. W., and A. M. A. Smith. 1987. Life History Parameters, Population Dynamics and the Management of Crocodilians. In Wildlife Management: Crocodiles and Alligators, ed. G. J. W. Webb, S. C. Manolis, and P. J. Whitehead, 199–210. Sydney, Australia: Surrey Beatty and Sons.

Weigel, R. 2014. Longevity of Crocodilians in Captivity. Int. Zoo News 61 (5): 363–373.

Weldon, P. J., B. Flachsbarth, and S. Schulz. 2008. Natural Products from the Integument of Nonavian Reptiles. Prod. Rep. 25:738–756.

Weldon, P. J., D. J. Swenson, J. K. Olson, and W. B. Brinkmeier. 1990. The American Alligator Detects Food Chemicals in Aquatic and Terrestrial Environments. Ethology 85:191–198.

Wiggins, M. 1990. They Made Their Own Law: Stories of Bolivar Peninsula. Houston: Rice University Press.

Wolfe, J. L., D. K. Bradshaw, and R. H. Chabreck. 1987. Alligator Feeding Habits: New Data and a Review. Northeast Gulf Sci. 9:1–8.

Woolford, B. C., and E. S. Quillin. 1966. The Story of the Witte Memorial Museum. San Antonio: San Antonio Museum Association.

Index

Other titles in the Gulf Coast Books series: